THE GLAMORG...
& ABERDARE CANALS
Past and Present

The Glamorganshire Canal in 1941. At top middle are the timber floats. Further down on the left is the set of factories of the Curran family. James Street Bridge is at the bottom left. Note the barrage balloons towards the bottom right. (Welsh Assembly Government Photographic Archive)

THE GLAMORGANSHIRE & ABERDARE CANALS
Past and Present

IVOR JONES

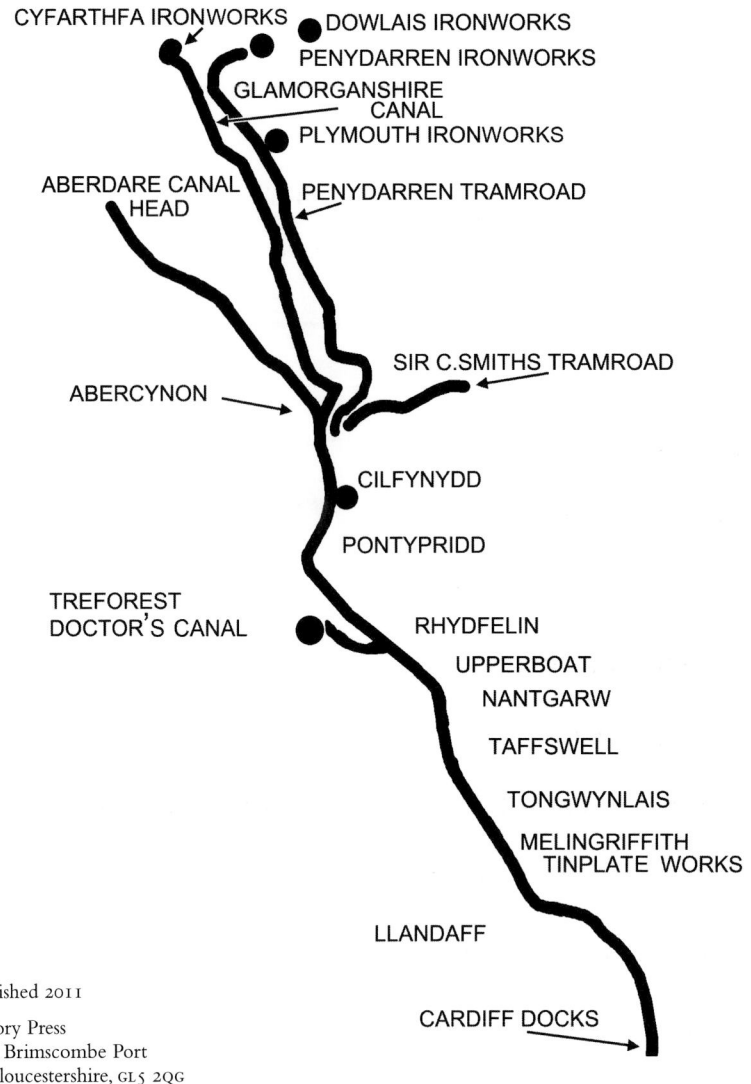

CYFARTHFA IRONWORKS   DOWLAIS IRONWORKS
PENYDARREN IRONWORKS
GLAMORGANSHIRE
CANAL
PLYMOUTH IRONWORKS

ABERDARE CANAL
HEAD   PENYDARREN TRAMROAD

SIR C.SMITHS TRAMROAD

ABERCYNON

CILFYNYDD

PONTYPRIDD

TREFOREST
DOCTOR'S CANAL   RHYDFELIN

UPPERBOAT

NANTGARW

TAFFSWELL

TONGWYNLAIS

MELINGRIFFITH
TINPLATE WORKS

LLANDAFF

CARDIFF DOCKS

First published 2011

The History Press
The Mill, Brimscombe Port
Stroud, Gloucestershire, GL5 2QG
www.thehistorypress.co.uk

© Ivor Jones, 2011

The right of Ivor Jones to be identified as the Author
of this work has been asserted in accordance with the
Copyrights, Designs and Patents Act 1988.

British Library Cataloguing in Publication Data.
A catalogue record for this book is available from the British Library.

ISBN 978 0 7524 6279 0

Typesetting and origination by The History Press
Printed in Great Britain

Contents

Introduction

On days off school, between the ages of approximately nine and fourteen, my friends and I would toss up for a decision on what to do that day. It was usually a choice between going to Heath Woods, Roath Park, the Wenallt, River Taff and Llandaff Baths, but mostly, it was 'the Canal'. We were a good gang and, as we grew older, we were more inclined to play cards, go to the Regal for snooker or the Plaza Cinema; and then, of course, the war came along.

I remember those canal walks very well but we never did see a working boat, except those at Cambrian Yard and Dock. We would walk to Gabalfa Lock via Western Avenue in its pristine newness and Professor Powsey would perform his travelling one-man show in the Lock Fields, climbing to the top of a lattice metal tower about 60ft high then diving into a round tank of water about 20ft in diameter and 6ft deep. This would have been around 1938–39. I also remember the lads that would dive off the bridge over the canal on the Western Avenue. If you threw a coin in, they would dive in and get it. The canal must have been deepened at that point because they certainly couldn't have done it in the 3ft depth of the rest of the canal!

These memories and many more lay dormant in my mind, until Steve Rowson and Ian L. Wright produced their wonderful volumes on the Glamorganshire and Aberdare canals. I first read them in the spring of 2009 and I have since gone over the ground that these gents did so well fifty years ago and enjoyed every minute. Most of the canals are well covered by roads and housing now but there are still many sites worth visiting today where parts of the canals, or the tramroads, can be seen. Rowson and Wright made the statement that they had 'opened up the research and dug out the facts and it was for others to elaborate and follow up' (or something along those lines!) and that is what I have also endeavoured to do. Hopefully my work will show the way to some of the spots of old industry in South Wales. In doing this job (if one can call it that) I have realised that this part of Wales is a very fair place to be and it has opened my eyes. Thanks to Stephen Rowson and Ian L. Wright for a good read.

Acknowledgements

Edgar Chappell, 'Historic Melingriffith' in *Old Whitchurch*

Edgar Chappell, *History of the Port of Cardiff*

Charles Hadfield, *The Canals of South Wales and the Border*, Cardiff, 1960

Stephen K. Jones 'The History of the Newbridge Works of Brown, Lenox and Company' in *Glamorgan Historian*, Vol.12

E.D. Lewis, 'The Pioneers of the Cardiff Coal Trade' in *Glamorgan Historian*, Vol.2

J.G. Owen, *The Pentyrch Iron Industry*

Don Powell, *Victorian Pontypridd and its Villages*

Richard Watson, *Rhondda Coal, Cardiff Gold*

Cardiff Central Library

Glamorgan Record Office

Llandaff North Library

Merthyr Tydfil Library

Pontypridd Library – Mr H. Matthews, Local Studies

Taffs Well Library

Whitchurch Library

Mr Brian Davies of Pontypridd Museum

Mr Medwyn Parry of the Royal Ancient and Historic Monuments of Wales

Mr Stephen Rowson and Mr Ian L. Wright

Councillor Mr Gordon Bunn

The Welsh Assembly Government of Wales, Photographic Archives – especially Mr D. Elliott; he and his team at the Archives have been superb.

1

The Glamorganshire Canal

By the 1760s it was widely known that there were considerable mineral deposits under the ground in the Taff Valley, especially in the north in the area of Merthyr Tydfil where there was everything needed to produce cast iron and then wrought iron, and even steel later on. There was iron ore, ironstone, limestone and coal for making coke. At this time nobody realised what the presence of coal would later mean to this iron-making area. Iron would play second fiddle to coal in the export market when the whole world was to buy the precious black stuff from South Wales.

In 1748 Lord Windsor leased some of his land at Dowlais to Thomas Morgan with mineral rights, and this gentleman, who already had a furnace in Caerphilly, declined the opportunity to search for the minerals that undoubtedly were there, and nine years later sold the lease to a partnership headed by Thomas Lewis of New House in Llanishen. This group had every intention of developing the site, as Thomas Lewis already owned works at Pentyrch and Caerphilly, and demand for iron was high. Abraham Derby of Coalbrookdale had developed a way of using coke instead of charcoal, and this new consortium at Dowlais wanted to go down that same road. By the end of 1760 they were producing 18 tons of pig iron a week. By 1767 output had doubled, and the company needed a man to run the works, as the men who had invested in the new company had other businesses to run. They brought in John Guest, who was to make the ironworks into the largest producer at Merthyr.

A few miles to the west of Dowlais the first of the ironmasters or speculators arrived in 1755. He was a gentleman from Whitehaven in Cumbria named Anthony Bacon who leased lands at Merthyr, a district 8 miles long and 4 miles wide, at the rate of £200 per annum for ninety-nine years, and built furnaces at Cyfarthfa in 1765. In 1777 Richard Crawshay of London became his partner, and in that same year Bacon acquired mineral rights at a place further down the vale, and he also gained the rights at Hirwaun. At both of these acquisitions furnaces were leased out, and then in 1783 he leased some of the Cyfarthfa properties to Francis Homfrey and his three sons, Thomas, Jeremiah and Samuel, who were to dispose of these leases to Richard Crawshay. Bacon died in 1786, and the Court of Chancery ruled that his Cyfarthfa property should be leased to Crawshay, who had already leased the remainder.

Bacon's property further down the valley was that of the Plymouth estate, and went to his former agent Richard Hill. His Hirwaun development went to Mr Glover of Abercarn.

John Guest died in 1785, and his son Thomas succeeded him, becoming associated with Mr William Taitt, who was to figure substantially in the Dowlais Ironworks, especially in his clashes with Crawshay about the canal later.

Meanwhile, when the Homfrays had given up their Cyfarthfa lease in 1784, they had taken over ground at Penydarren and built ironworks there, of which Samuel Homfray became sole manager in 1789.

These four works, Cyfarthfa, Dowlais, Penydarren and Plymouth, went into production of iron, and the substitution of local coal for charcoal in the smelting process at Cyfarthfa and Penydarren about 1787 ensured a massive increase in iron production.

The iron ore, coal and limestone were to be obtained locally, but the finished product could not be delivered economically to the place of export, at Cardiff, as there were no railways or good roads. Anthony Bacon used mule trains to Cardiff and Swansea, and he pioneered a route from Merthyr, through Gelligaer and Caerphilly, to Cardiff – improving on an old road about 1767.

A Turnpike Act was produced, to enable a turnpike road from Merthyr south to Tongwynlais, where it would join the Cardiff District Turnpike. The Act had trustees Jeremiah and Samuel Homfray, and James Harford of Melingriffith.

Even on these improved roads a wagon could carry only 2 tons of iron to the old quay in Cardiff (Quay Street), so it was too expensive, costing the ironmasters £14,000 a year.

Led by Richard Crawshay, the men connected to the four ironworks at Merthyr, together with prominent Brecon people, including the proprietors of Wilkins Bank, the owners of the Melingriffith Tinplate Works and some backers at Cardiff, joined together to obtain the Act of 1790 to build the Glamorganshire Canal, from Merthyr Tydfil to Pontypridd and Melingriffith, to the Bank – a shipping place on the Taff below the town quay.

It was to be the first canal of any substantial size in Wales, with an authorised capital of £60,000, and with the potential to raise £30,000 more, with a limit of 8 per cent on dividends. A branch canal to Dowlais from Merthyr was also to be built, but a survey revealed that Dowlais's elevation up on the side of the valley would mean 411ft of lockage in only 1¾ miles, to cost £16,282, in addition to the main canal cost of £53,465. This idea was dropped by the committee, to the consternation of the Dowlais board.

The shareholders were Richard Crawshay who put in £9,600 and his family who subscribed a further £3,500. William Stevens gave £5,000, the Harfords of Melingriffith £6,000, John Kemys Tynte £5,000, the Homfrays £1,500, Richard Forman of Penydarren £1,000, the Hills of Plymouth £1,500, William Taitt £1,000 and Thomas Guest £500. At their first committee meeting at Cardiff they decided to engage Thomas Dadford senior, Thomas Dadford junior and Thomas Sheasby as joint contractors to make the Glamorganshire Canal for £48,288, exclusive of land purchase. A bond of £10,000 was taken from them for the finishing of the job.

The canal was to be 30ft wide at the surface of the water, and 4ft deep. The bridges were to span 18ft, with a haling path under them.

The locks were to be 60ft in length from gate to gate, and 9ft 6in wide.

After the survey, the route from Merthyr was to travel down the west side of the valley, on the hill slopes high above the Taff for 9 miles, and upon reaching Abercynon would change to the other side (east) of the valley for 15 miles to Cardiff.

The Glamorganshire Canal Company's (GCC) agents were busy buying up land along its length, and the workforce arrived. They were Irish, Cornish, Welsh and many other nationalities, and included masons, carpenters, quarry men, sawyers and navigators. What a bunch of hard men they were. They had worked on many other canals and docks, railways and roads. They worked hard and lived hard, rarely seeing home. Work on the route reached Merthyr in August 1790.

The work started at Cyfarthfa, and already the fact that Richard Crawshay was the senior shareholder began to show; it was his works they served first. He wanted an extension built from the originally agreed canal head to a place further north to the Cyfarthfa Ironworks, in 1791. Half a mile of extra cost and further delay, and right from the start a quarrel had developed between Crawshay – who had virtually a controlling interest – and Taitt of Dowlais. On the occasion of the dropping of the Dowlais canal branch, Crawshay told Taitt that the Dowlais Company must find its own way to the canal. At the same time the maximum proposed toll of *3d* per ton per mile on which the capital had been subscribed was raised to *5d*. Taitt protested that the Dowlais Company could send, as they had been doing up to that point, their iron over hills on horseback as cheaply as they could on the canal. Taitt therefore declared that he would have nothing more to do with the canal, and neither he nor Guest was on the first committee. However, Crawshay compromised, and a tramroad was built by the Dowlais Company past the nearby Penydarren Works to the canal at Merthyr, costing about £3,000, to which the canal company contributed £1,000 in June 1791. Almost immediately Homfray and Crawshay also quarrelled over the running of the canal, and Homfray also argued with Taitt; when Taitt built the new tramroad from Dowlais to the canal, he had bypassed Homfray's works at Penydarren, and the argument was that Taitt could have brought the line of the tramroad out to pick up the Penydarren traffic on its way to the wharf. Taitt wrote:

> Your several letters to Mr Guest about the direction of our railroad from Dowlais to the canal surprises [sic] me exceedingly as I know not what pretence you can possibly have to interfere in that … If you want an accommodation from us, ask it as a favour, but do not think of demanding it as a right.

This resulted in Forman of Penydarren leaving the canal committee, so that only two Merthyr works were represented (until Taitt rejoined in 1795) for three years. What a turbulent time, yet one that would lead to a wealth that none of them realised was coming their way.

There was another extension planned at the southern end of the canal, from the original idea of finishing at the Bank on the Taff just south of the quay in Cardiff, to a point further south to the mouth of the river, a mile further to the sea. It was clear that Dadford was going to need more money than the original estimate. He produced a bill for £17,220 for the extra work, and he stated that a further £5,000 needed to be spent. The basin at the sea lock was to have a depth of 16ft of water, and was to be not less than 40 yards wide; the sea lock itself was not to be less than 30ft wide and 90ft long, and the entrance to the lock not less than 36ft wide.

The canal was opened on 10 February 1794, and there was a fleet of boats at Cardiff, all laden with the produce of the Merthyr Ironworks. Richard Griffiths of Cardiff, a committee man (also of the Doctor's Canal), gave a show of entertainment and celebrations for which the company paid £14 11s 9d. The canal tolls on this day were 2d per ton per mile for coal, ironstone, iron ore, limestone, lime, manure, bricks, clay and sand, while the maximum 5d per ton per mile was charged for iron, timber, goods and merchandise.

The canal was now 25½ miles long with forty-nine locks along its length, with later improvements extending to fifty-one locks. The rise from sea level to the canal head at Cyfarthfa was an astonishing 543ft. The main engineering work was at the stone aqueduct at Abercynon, which spanned the River Taff (now a road bridge) and the 115-yard tunnel under Queen Street in Cardiff. The total cost, including extensions and basin, was £103,600.

Just as everyone was rejoicing, a row broke out between the contractors and the GCC after completion of the canal in 1794. A breach of the bank occurred in December and Dadford and Sheasby refused to repair it unless the company advanced money for it. This was not done, and the contractors had not only finished the canal, but had been running its daily affairs, so they fired all the men and walked away. The GCC sued the contractors, and they were arrested. At the trial, an arbitrator was selected, Robert Whitworth – a notable canal engineer – who in the end found for the contractors, not the GCC. This was something of a surprise for Richard Crawshay, the major shareholder and chairman of the GCC. He was a bit of a tyrant, a capitalist of the old school, and he managed to alienate the other ironmasters by putting pressure on freighters and boatmen to give his goods priority in the queue of work to proceed to Cardiff and the return.

The next controversy occurred when Richard Hill complained that the canal was taking water from the Taff that was legally his, and in 1797 there was litigation between the Dowlais Company and the GCC over the number of pounds to be allowed in the hundredweight. Richard Crawshay was behind this business and he ruled over the other committee men, trying to force other ironmasters to comply with his wishes and needs, which usually meant that Dowlais and Penydarren works suffered, so the two bosses – Richard Hill of Plymouth and William Taitt of Dowlais, with George Overton as engineer – built the tramroad from Merthyr to Quakers Yard. This line opened in 1802 as a 4ft 2in-gauge plateway 9½ miles long from a junction with the existing Dowlais tramroad to the east of the aqueduct at Abercynon. The Penydarren tramroad was owned in the proportion of five shares each to the Dowlais and Penydarren companies, and four shares to Plymouth. This is the line that was to become famous in 1804 as the one that carried Richard Trevithick's steam engine on the world's first rail journey powered by a locomotive.

Dowlais ceased to use this road when the Dowlais railway was built to connect the works with the Taff Vale Railway in 1851 and they transferred their shares to the TVR.

The Penydarren Works closed in 1859, and the remaining owner, the Plymouth Works, gave up making iron in 1880. The tramroad was converted into a railway between Merthyr and Mount Pleasant, and the rest abandoned.

The Penydarren tramroad was built as a result of Crawshay's quarrels over canal tolls and charges, as well as his taking water from anyone else who could not, or would not, resist him.

Water, or the shortage of it, caused a great deal of stress and many legal challenges over the years. The GCC was forced to bring in Mr Rennie, an expert on the matter, in 1806, and as a result a reservoir was built at Glyndyrys. Also installed was a Boulton and Watt pumping engine at Pontyrin which was to pump back to the canal the water that had passed through Hill's works. A waterwheel and pump was installed at the tail race of the Melingriffith Tinplate Works. To anybody unfamiliar with these old industrial works it may be useful to learn that water was not needed for the processing, it was needed for its power. It was to turn large turbines, then the rolls and presses, and until the use of steam engines, the shortage of a sufficient head of water could close down a factory. The water was obviously short in the summer, but the works needed it in all seasons. The canal was a rival for that water, and before the canal was built the ironworks was using the water that was available. The illegal use of the Taff and other waterways maintained a depth of water in the canal that was consistent with passage of a boat loaded with 20 tons of cargo. Yet another reservoir was built below the Treble Locks at Glanyllyn, in an attempt to improve the supply at Pentyrch Ironworks, and Melingriffith, but it was not until 1829, after years of argument and litigation, that an agreement was reached.

The dispute was to break out again in 1832 and last for some years, until it finally ended. Richard Blakemore, who had been associated with Melingriffith since 1807, and who had taken it over in 1812, was said to have spent £20,000 in the courts over thirty years.

TAFF VALLEY IRON PRODUCTION

Richard Crawshay sat on the GCC committee until his death in 1810. His son William Crawshay had had a seat on the board since 1798. The Cyfarthfa Works in 1815 had seven furnaces and made 18,200 tons of iron, at an average of 50 tons weekly. In 1845 this had increased to eleven furnaces, with an output of 45,760 tons of iron, at an average of 80 tons weekly. This success led to the opening of a new mill at Cyfarthfa on 18 March 1846. In this new rail mill in March 1847 the enormous amount of 1,144 tons of railway iron was produced in just a week.

Dowlais Ironworks in 1806 produced 5,532 tons of iron. In 1815 its six furnaces made 15,600 tons at the average of 50 tons weekly. By 1845 the works had eighteen furnaces and made 74,880 tons, an average of 80 tons per week. At Penydarren, the quantity of iron sent to the canal from October 1805 to October 1806 was 6,963 tons, under the ownership of Mr Homfray. In 1815 Penydarren had three furnaces and made 7,800 tons – figures that doubled by 1845, when 15,600 tons were made (80 tons per week) in six furnaces. The Plymouth Works, belonging to Anthony Hill, sent out 3,952 tons of iron in 1806. In 1815 they had three furnaces and made 7,800 tons. In 1845 they had seven furnaces and made 29,120 tons at the average of 80 tons weekly.

◆ ◆ ◆

The competition expected from the new Penydarren tramroad did not depress the shares of the GCC; in fact there was an embarrassment of riches. In 1805 the committee, bound by the limit of 8 per cent dividend by Parliament, could not get rid of surplus revenue, so it was decided to return 20 per cent of the tolls received back to the traders. This gesture had the result of attracting more freight, so money was quickly spent on improvements to the canal, especially to the basin at Cardiff. Tolls were reduced to carriers in 1815 by 10 per cent, which increased traffic even more, so nine months later it was decided to charge traders no tolls at all for the last quarter of the year, and again for the first quarter of 1817.

By September 1818, 50 per cent reductions in tolls were the norm. The year 1820 saw a further 5 per cent reduction, and in June 1821 £3,648 was returned to the traders. Unsurprisingly, this act went down well with the traders and the ironmasters whose cargoes were being moved to Cardiff, and in the opposite direction as there were tons of foodstuffs, grain, shop goods and the like being brought up. However, the 'feel-good factor' did not extend to the canal workers. There were no wage increases for them; in fact a meeting of the GCC asked Crawshay and the clerk to make such wage reductions among all the staff as would seem consistent with the honour and liberality of a public body. The GCC found further ways to reduce costs: in March 1822 Crawshay reported that he had reduced the wages of the canal staff, but the committee ordered that, 'In future no man to be taken into employ above the age of thirty, and that no tools or beer be allowed the men.'

It was a capitalistic horror. It was not as if the men were being paid well. They were made to work hard in all weathers, and by day or night. Before the wage reductions the ruling wages were 12s a week for labourers and up to 18s a week for a mason. Is it any wonder that socialism was to flourish here?

The real reason was that all the committee men were ironmasters that were employing thousands of men, women and children, all working for a pittance in very harsh conditions, and if they paid the canal workers any more than the iron workers then they would be pressed to pay the iron workers more. Between 1922 and 1924 further work was done at Cardiff. A total of £3,600 was spent above the sea lock pound, and it enabled boats to increase their capacity from 20 tons to 25 tons.

This work was probably suggested by George Overton who had been brought in at that time to suggest a cure to a serious problem. The floor of the canal, and the berths along the wharves, were seriously affected by silting, and some ships were still dumping their ballast secretly.

In October 1828 a further toll reduction of 5 per cent was made, making a 75 per cent reduction below parliamentary rates. The Glamorganshire was the richest canal in the whole country, which resulted in another reduction of 2½ per cent in 1830 – a significant year, as the Bill for Lord Bute's West Dock was passed, and this year saw the canal carry 201,116 tons of iron and coal. Of this total 87,376 tons of iron were shipped as follows: Dowlais 29,621 tons; Crawshay's 21,312 tons; R. & A. Hill 13,046 tons; Penydarren 12,246 tons; Aberdare Iron 7,248 tons; Blakemore 2,894 tons; S. Brown 664 tons.

The total amount of coal shipped was 113,749 tons, the largest parts by: Walter Coffin, 46,446 tons; Sir C. Smith, 18,246 tons; T. & G. Thomas, 11,400 tons; Robert Thomas, 10,476 tons.

The great era of steam coal exports began at about this time. From this period the docks became busier, and at canal head tolls had been reduced a further 10 per cent, making it 85 per cent lower than parliamentary maximum allowed. It now was the most prosperous canal ever.

In 1839 the Bute West Dock opened, complete with a junction with the canal at its northern end.

Anthony Hill, of the Plymouth Works, was fed up with the increasing hold-ups, from various causes, and was keen on having a railway at their convenience, and he also knew Isambard Kingdom Brunel. He asked the famed railway builder if it was feasible to build a railway from Merthyr to Cardiff, and he also involved J.J. Guest of Dowlais, who had proposed a railway some time earlier, not to Cardiff but to England, which would have had the name Cambrian, Gloucester and London. This never happened, but something had to be done, as the difficulties in congestion worsened, even after day and night shifts were worked at the locks. The number of men doubled at the locks, and oil or gas lamps were installed at each lock to prevent delays becoming the norm.

To save water at the locks, they adopted a system whereby 'working turns' were started; this meant that boats going up waited for a boat coming down, so that a lock-full of water would pass two boats.

Making matters worse was the fact that the ironworks were now importing iron ore, as the local ore was both getting scarce and reducing in quality, so the boats had to bring ore from abroad up from Cardiff. For example, the Dowlais Company sent down 39,000 tons of iron from Abercynon in 1835, against only 24,000 tons in 1831, but imported 15,668 tons of iron ore, against 6,156 tons in 1827. These boats heading north with 20-plus tons on board had to overcome a pronounced current generated by dozens of lock-loads of water coming downhill. The busier the locks, the worse the current. All this, plus the fact that the canal above Abercynon was getting shallower. For these reasons the boats had to be lightened, and lighter boats meant more trips. Things were going wrong with the wealthy canal.

This was why Hill and the Guests pushed more ironmasters to form a railway. At this time coal was not the factor it would become. It was still iron that sustained the canal, but times were changing.

On 12 October 1835, at a meeting in Merthyr, the Taff Vale Railway was formally inaugurated, with the Hills, Guests, Walter Coffin and a member of the Charles family on the provisional committee. The Taff Vale Railway got its Act of Parliament in 1836.

The railway was opened to Navigation House, Abercynon, from Cardiff in October 1840, and to Merthyr in April 1841, but the anticipated loss of business on the canal, due to freighters decamping to the railway, didn't happen because, at around the same time, there was a fortuitous massive increase in domestic sales and exports of coal. Every system of transport was up to their necks in work. Ten years after the Taff Vale began its work the following figures are worth noting:

IRON AND COAL TRANSPORT, 1851

	Iron	Iron Ore	Coal
Glamorganshire Canal	190,633 tons	94,408 tons	294,537 tons
Taff Vale Railway	74,701 tons	51,000 tons	580,000 tons

However, the canal's tonnage fell from 466,985 tons in 1858 to 315,749 tons in 1868. There were many causes, including the canal-side works and collieries connected up to the railway and the opening of the Bute East Dock, and that dock being connected to the Rhymney Railway in 1857. The canal company dug a junction canal through to this new dock as they did to the West Dock in 1859.

By 1865 the canal head at Cyfarthfa was no longer in use on a regular basis, and all work was being done south of Jackson's Bridge. The piece of real estate from this bridge north to Cyfarthfa and canal head was owned by the GCC, then when the canal was purchased by the third Marquess of Bute a deal was made to give this northern end to the Crawshay family in exchange for the Crawshay lands on the banks of the ship canal at Cardiff, to enable the Marquess of Bute to build timber ponds on the site; this was done in 1885.

The fall in freight continued; the canal's traffic had reached its peak in 1851 when 580,000 tons were carried. Amazingly this was ten years after the Taff Vale Railway had opened in competition with the canal, showing how rich were the pickings in the coal and iron industries, but inevitably the competition began to have its effect.

In 1866 the Bute Trustees offered to buy the GCC, but were refused.

In 1882 the company's dividend was down to 1½ per cent. The 8 per cent dividend paid almost from the beginning could not be maintained after 1876, and the company was authorised to make a short railway branch from the sea lock pound. In 1883 work was begun after it was decided to borrow up to £27,500 on mortgages.

At this point the Bute estate made another offer to buy all the canal shares, and this time it was accepted on 19 November 1883. The benefits for Bute were clear: the GCC could no longer build its own dock, as they had been attempting to, or interfere with his own docks bills at Parliament, as he attempted to release water for his docks and perhaps to bargain with the Taff Vale Railway. As a result of his acquiring the GCC, the Taff Vale reduced their rates on coal for shipment.

Lord Bute took over the canal in 1885, and formed a new committee with himself as chairman, and the name of Crawshay finally disappeared from the records. The new chairman called for a report on the status of the canal, and this resulted in a gloomy picture of decay. The locks and buildings were in a poor state. Of the whole property from Cyfarthfa to the sea lock gates there was not a single house which would not be improved by a mason, carpenter or painter. It was resolved to put the canal in repair, and to make good subsidence damage. The new owner reconstructed the sea lock, enlarged and deepened the basin, and provided timber ponds. From 1887 onwards, no dividends were paid, and in 1888 it was decided to promote a Bill to convert the greater part of the canal into a railway, because in that year James Abernathy and G.B. Bruce had reported that for all purposes of traffic, except on a portion near its terminus, the canal was obsolete.

In 1896 a breach at Cilfynedd occurred and was repaired, but when another breach occurred there in 1915, it was not repaired, and the canal above Cilfynedd was closed.

In 1897 the Bute Docks Company became the Cardiff Railway Company, but the line that was built never used the line of the old canal bed, and it only reached Treforest after passing through the last station at Rhydyfelin. Failure to gain access to the Taff Vale lines was the end of Lord Bute's dream of sharing the freight fortune from the coalfields.

◆ ◆ ◆

The canal was only carrying a small proportion of the total freight coming down to Cardiff, and in 1898 and 1889 two steam-driven boats with funnels that hinged down flat to go under the bridges were introduced, but they were unable to show their real value because of the number of locks.

In October 1908 a serious breach in the canal wall at Dynea occurred. This was a site of repeated flooding, as the locality still is to this day. This was repaired, but the breach further north at Cilfynydd in 1915 was not. The topmost section of the canal from Cyfarthfa to Merthyr had been disused since 1865. The part from Abercynon to Merthyr was closed on 7 December 1898, and was sold to Cardiff Corporation to be adapted as a water main. From then on the canal became a water feeder. The breach at Cilfynydd caused the closure of the Abercynon to Pontypridd stretch, but the canal continued to be used between Trallwyn and the sea lock, with a few old customers like the Melingriffith Tinplate Works attempting to carry on as if nothing had changed, until the end came on 25 May 1942. A breach occurred at Nantgarw, effectively closing the canal except for a small part of the sea lock pound. In 1943, Cardiff Corporation bought the canal, and an Act of that year authorised its abandonment.

The sea lock pound went on serving light coastal vessels, two sand dredging companies' ships, the occasional Neal & West trawler and pleasure boats. The wharves carried the engineering works of several small repair firms, when again disaster struck on 5 December 1951. One of the sand dredgers accidentally drove into the lock gates of the sea lock, emptying the water from the pound.

Cardiff Corporation was pleased as they had been trying to close the dock down since 1942. The two dredging companies were forced to find other docks. Bowles Sand & Gravel went to a Taff-side tidal berth between Clarence Bridge and the mouth of the Taff. Tuckers Sand Dredging went to the Bute West Dock, along with the small coastal ships.

◆ ◆ ◆

One of the things that had kept the canal going profitably from Abercynon to Cardiff was the new fuel block trade. In 1858 Mr T.H. Wood built the Crown Preserved Coal Works at Llys Talybont near Maindy, Cardiff. Then in 1866 the Anchor Fuel Works opened, followed by the Star Works in 1872. All these works, with slight differences in process, made blocks of small coal and pitch, hot pressed into moulds. Each firm had their own dimensions and logos, so that there was no confusion which was which. The factories were along the canal bank at Maindy, and another, fourth, company – the Cambrian Fuel Works – opened at Gabalfa.

All of these works loaded their blocks on to canal boats at their respective wharves and were towed to the pound at Cardiff, either to be unloaded there or to go through the junction lock into the West Dock. This latter option was enforced when the Marquess of Bute gave notice to quit to the companies of freighters along the wharves of the pound.

This trade was really taking off. Shipping companies used the blocks for the bunkers of their ships, and as the blocks were composed of useless small coal it was readily available from collieries, but it had to be the best steam coal.

In 1898 the Crown Works left its canal-side location and moved to a more modern, larger plant at the Roath basin. Then the Cambrian Works closed down and became the headquarters and central workshops of the GCC that was at Navigation from the beginning. The Star and Anchor works continued to sell their fuel blocks until the late 1920s and early 1930s. The sites that they occupied are now university student domestic blocks, and except for the occasional change in the shape of the now filled canal bed, it is hard to realise that such a noisy, filthy place could have been here at this peaceful, green and wooded place.

This photograph of 1910 shows the water supply to the canal head (at the bottom of the shot). The railway is the Crawshay private line for the Cyfarthfa Ironworks. In the middle of the picture is an old lime kiln, and above that, the River Taff, hidden in its gorge. At the right is Pandy Farm. Cyfarthfa Castle is at the upper left. This waterway was one of several feeding the upper end of the Glamorganshire Canal – but there was never enough water to do the job. (Merthyr Library)

RAF 16042 of 7 March 1945, depicting the area of the canal head forty years after the closure of this northern end. The head is marked with a cross just south of Pandy Bridge. Following the canal southward we see the railway, shown in the previous photo, emerging from a tunnel then passing under Pandy Bridge and running alongside the wharf at the head's west side. Lower down on the canal's west side are the GCC's workshops, then the space of canal-side that was the graving dock – now filled in.

Jackson's Bridge is at the bottom left and the west side of Lock No.1 can be seen standing above the bridge. On the east side of the canal, going south from the head, is the rail finishing shed, then Crawshay's Chapel of Rest and Chapel Row. The GCC's stores buildings are still evident but the basin has been filled in, as have the canal and docks. (Welsh Assembly Government Photographic Archive)

This is just about all the tangible evidence that is left showing that the canal ever existed in Merthyr. Chapel Row and the Chapel of Rest are on the right. We can see the short stretch of reconstituted canal and the Rhyd-y-Car cast-iron bridge that was moved to this site post-war. Drams and rail are also tastefully displayed, but it would have been nice to see it continue to the canal head, just 200m further north. (I. Jones)

Jackson's Bridge today, with no sign of the canal on the far side of the River Taff. The entire area of the canal lies under modern housing and roads. (I. Jones)

This photograph shows Lock No.1 of the Glamorganshire Canal near Jackson's Bridge. The company's warehouse is on the right and Cyfarthfa Castle is on the horizon. (Merthyr Library)

Penry Street Bridge was the location of the Middle Lock (on the far side of the bridge). This bridge carried the road from Swansea into Merthyr. Today nothing is left of this place, except the new river bridge replacing the iron bridge, and there is no canal bridge. (Merthyr Library)

The superbly refurbished old engine house of Ynysfach Ironworks near the line of the canal. (I. Jones)

Rhyd-y-Car Bridge, near Ynysfach Ironworks, is one of the only surviving bridges at the top end of the canal. This bridge has been spared from the spoil tip bulldozing that happened when the Ynysfach tips were spread. Nevertheless, this old bridge is in a crater surrounded by spoil. (I. Jones)

2

Down to Abercynon

THE DOCTOR'S CANAL

A strange name for a canal, but the man inspired enough to build it was a qualified doctor of medicine who had a practice in Cardiff along with a partner, Dr Richard Reece. The man in question was named Dr Richard Griffiths, and he was to become a key figure in Welsh industrial history; in addition to his other interests he was the first man to send Rhondda coal to the outside world. Born in 1756, he took a lease of the minerals under the Hafod Uchaf lands at the end of the eighteenth century.

After spending his early years as a doctor, he lost focus and became more interested in gambling and the fascinating Klondike-style atmosphere growing rapidly in the ironworks, mining, tramroad building and canals. He was eager to join this frenzy of investment.

He became a committee member of the GCC in 1793, and that was when he bought the leases of mineral-bearing lands, with a view to selling the leases to others at a much higher rate, when the sites were proven.

In 1807 he asked the GCC whether they would help him build a tramroad from his colliery at Gyffeillon to their canal by giving him £3,000 of free tolls on his first coal shipments, but they refused. A year later he proposed a canal branch a mile long from the Glamorganshire Canal, at Dynea, to Treforest, and a tramroad from there to a bridge over the Taff, and then on for 3¼ miles to Hafod. This was agreed, and when he further asked that the building materials for the project be carried free by the canal company, that too was agreed. The canal company could obviously see that profits could be made from this freighting of coal.

The tramroad and river bridge were built in 1809, under the 'four-mile clause' of the canal company which allowed any business within 4 miles of the canal wishing to transport their goods to build their tramroad or canal without applying to Parliament for permission to buy land to build over. Dr Griffiths had built his tramroads from Cwm Rhondda and Hafod to connect to his canal-to-be, so he began to use the basin and wharf alongside the Glamorganshire Canal at Pwllywhyad Bridge, no more than 200 yards from his proposed canal head that lay to the west of the bridge. It only required laying rails of the tramroads to veer eastward and climb a slope. Coal was shipped from here between 1808 and the opening of the Doctor's Canal in 1813. Some fifty years later coal from Bryn Tail mine was shipped from here, brought down by tramroad from the other side of the valley.

It was not until 1813 that the GCC would allow him to connect his canal to theirs, because they were not satisfied with his water supply. This was settled and by then Mr Walter Coffin had built a tramroad from his collieries further up the Rhondda to connect with the doctor's tramroad in 1812. This was steam coal, and was a much sought-after freight.

The new canal attracted many customers, and wharves were constructed along its length from firms that had no other outlet for the trip to Cardiff.

The Doctor's Canal supported two boatyards along its mile; one at Dyffryn was used by John James with three men working for him, and by the turn of the century he had moved his boatbuilding and repairing yard down to Gabalfa at Cardiff. The other boatyard was at Dynea just north of the point where the two canals joined. This yard was called Blain-y-Llyn, and was first in the hands of Ebenezer Morris in 1841, and later by David Jones who employed four men. He built cottages for his workers across the canal from his boatyard.

Dr Griffiths died in 1826 aged seventy, leaving his property to his nephews, the Revd George Thomas of Ystrad-Mynach and his brother, Thomas Thomas of Pencerig, Radnor. They both continued to operate the tramroad and canal, so that is how the canal now became Thomas's Canal. This occasionally causes moments of confusion but it is still the Doctor's Canal. He had transformed the Treforest area on his own, and the town developed faster than Newbridge (Pontypridd). The tinplate works on the west bank of the Taff and the collieries kept the boats loaded for the trip to Cardiff, and foodstuffs and other shop goods came up. The canal began to lose its customers once the Taff Vale Railway opened, and there was only a small boat trade by 1904, which ceased by 1914. By 1918 it was derelict. Nothing remains of the canal today, except for the fact that the roads of housing at Rhydyfelin follow the line of this private venture.

THE BRYN TAIL COLLIERY

Up high on the slopes of Mynedd Eglwysilan, coal was found on a level south of Bryn Tail farmhouse, the childhood home of William Edwards – the builder of the famous stone hump-back bridge in Pontypridd – but later the home of the notorious Dr William Price, the eccentric druid who introduced cremation to this country when his son died. He was the owner of the land (or so it was believed), as well as a mineral lease, dated 2 February 1807, of the 115-acre Bryn Tail estate, then owned by John Richards of Llandaff. The twenty-one-year lease was to Samuel Harford, Richard Jones Tomlinson, John Harford and Richard Blakemore; these were the names of the shareholders of the Melingriffith Works. This is strange as this company had several coal suppliers much nearer to Cardiff at Graig-yr-Allt and Garth. No evidence has come to light of coal working being attempted at this time.

Doctor's Canal head about 1947. This photograph was taken before the place was levelled and built on. The Rhondda coal came down the miles of tramroads, over the Taff on Machine Bridge, and south to this place which had wharves on both sides. Note the lamp post on the left, on the path that would have taken you to the front of the terrace, facing Pwllwhyad Bridge and Wharf. Behind the camera, when the canal was active, lay a busy tramroad marshalling yard with rail track going all about the area; two tracks also led off to Pwllwhyad Wharf. (Cardiff Central Library/ Ian L. Wright photo)

At the foreground of this photograph was the junction. Mr Bill Pain, who lives at one of the houses on the left, is old enough to remember the Doctor's Canal and has studied its past. He was, and is, a great help to me. (I. Jones)

Whilst installing flood-prevention measures in the 1980s, the local water and environment people dug away on the spot of the Doctor's Canal. They unintentionally revealed some of the 200-year-old masonry work. (Welsh Assembly Government Photograph Archive)

An oblique view of the Treforest Trading Estate during the Second World War, taken during an RAF sortie of 21 April 1943, showing many places mentioned in this book. At the top left marked (1) is Maesbach Farm; (2) is Maesmawr Farm; (3) is the Dihwyd tramroad track and incline (dotted); (4) is the coal level with Warren Farm to the left; (5) is the main railway line (first Taff Vale Railway then Great Western Railway); (6) at the lower left is Maesmawr Bridge on the canal; (7) the Cardiff Railway track; (8) the Glamorganshire Canal; (9) at the bottom left is a glimpse of what was the Alexandra Dock and Railway line. The old estate power station, at the centre, was demolished and replaced further upriver. The accumulated cinders and ash from this demolition were used to fill the defunct Doctor's Canal. The '+' at upper left shows where George Insole sank his shafts by the Taff and the '+' on the right shows the coal levels on Maesmawr land, now difficult to trace in the woods; (10) is my approximation of where the unique Maesmawr coal ferry operated. (Welsh Assembly Government Photograph Archive)

Dr William Price was the surgeon at the Brown Lenox Works, and in November 1857 he purchased and mortgaged the adjoining estates of Bryn Tail and Craig Alfa from John Rogers the younger, who had inherited the freeholds from the Revd Evan Jones in 1835. In May 1858 the 4ft Gwenhiolen seam was stuck only 16 yards from the surface. With the mining engineer's report he was ready to lease the mineral rights. William Harris and Charles Davies junior, as partners, took over the coal level and a fourteen-year term was granted them in the colliery lease of 17 June 1859. Rent would be £500 per annum, together with 1s per ton for all coal brought out in any one year above 10,000 tons.

To operate this site and run a tramroad down the mountain, it was necessary to buy land to build it on; that meant some of Dr Price's land, some of Baroness Windsor's land and some of Lord Llanover's to build the inclined plane to the canal below. These people, including Price, would not or could not allow this. Price lost interest and fled to France. Meanwhile, the worried Harris and Davies used the delay to develop the level workings toward the brow of Eglwysilan where the seam thickened. By December 1859, with no prospect of Price furnishing them with the land for the incline, they had no means of transporting the coal to market, and so started the slow process of litigation. In December 1860 the lessees filed a Bill in Chancery against William Price and his mortgagees to compel him to provide the land, claiming costs, and reserving rents and royalties due under the lease, until the case was resolved. In his absence abroad, Price arranged for his solicitor to be assigned his mortgage of Bryn Tail and Craig Alfa. While the legal battle to recover Harris and Davies' costs continued, the estates were sold by auction on 16 August 1862. The two speculators filed for bankruptcy, as they were never able to recover costs because Price claimed immunity under the Statute of Limitations Act by living in France for seven years. At the auction Charles Davies junior's father backed his son and purchased the estate from Price's solicitor, and the old workings at the level, dormant for three years, were put into shape. By March 1864 the incline was complete, and on Wednesday 4 May the colliery was opened.

It was an 800-yard incline, including a horse-drawn tramway for the last 300 yards down to the basin and bridge at Pwllyhwyad on the Glamorganshire Canal at Gwern-y-Gerin, and at last the coal was being delivered. Five months later the owner Charles Davies senior was advertising for sale the whole of Bryn Tail estate and Craig Alfa Quarries. It is a strange story, that after sinking so much money in the infrastructure and all the frustration that came with this mine, when he was finally achieving some returns he should sell. It is not known if a buyer was found, but it is significant that all the other collieries were seeking to connect up to the Taff Vale Railway, and there was no chance of that happening to this small property high up on the mountainside. An Ordnance Survey map of 1875 shows Bryn Tail Colliery and incline as disused.

The signs are still there of the tramroad, though the bridge was demolished. When Glyn Taff Crematorium was built the 300 yards of horse-drawn tramway was obliterated by the development. The whole place is worth a visit, but you do have to be persistent to find any remains.

A TRIP THROUGH PONTYPRIDD ON A CANAL BOAT

The canal south of Abercynon, to the outskirts of Newbridge (Pontypridd) went through open fields until, from 1886, the villages of Cilfynydd, Pontshonnorton, Coedpenmaen and Trallwn sprang up after the sinking of the Albion Colliery at Cilfynydd in 1884. This stretch of canal ran across the meadows of Ynysydwr and Craig Evan Leyshon Common, and the boatman could enjoy the peace and quiet. The sides of the canal were bounded by cornfields riddled with scarlet poppies and blue cornflowers as the boat floated towards Cilfynydd Inn, seen in the distance. The canal enters Cilfynydd at Lock Cottage, where the canal is stepped down by Ynyscaedugwyd Lock, No.25, to pass under a small hump-back bridge which gave access to and from the Albion Colliery.

At Cilfynydd the boat horses were often stabled here, in what became Cwm Cottages, where Cilfynydd Rugby Football Club social club is now sited. Leaving for the south, the canal reaches quarry sidings, where many canal horses were shod at the farrier's and blacksmith's shops, on a site where later Kings Garage was sited, and boats would load up with stone from the depot belonging to Bodwenarth Quarries. Continuing southwards, we arrive at the stone bridge over the canal at Pontshonnorton (John Norton's Bridge) where the boats might load with coal from the Bodwenarth Colliery, sunk in the early 1870s.

The canal then ran down to the bridge at the foot of Coronation Terrace. The name of the bridge became Distillery Bridge because it gave access to Chivers' distillery and chemical works, which dispatched much of its production by boat. The canal then runs parallel with Coedpenmaen road to a lock, opposite the Baptist chapel in the village of Coedpenmaen – this is Tynygraig Lock, No.26. Then on to the busy trading and commercial part of the canal, dwarfed only by the sea lock pound at Cardiff. The next lock, and the bridge carrying the main Coedpenmaen road, is lock No.27, known as Road Lock, with the eighteenth-century Newbridge Arms pub facing it. The pub is still there, but the lock, basin and boatbuilding yard are gone, under the western edge of the A470.

The next stop on the canal is also a part of the busiest section of the entire canal, except for Cardiff. It is Trallwn, and its locks and wharves, a location where the goods arrived from all over the place. First is Lock Odin, a double (Nos 28 and 29), then the canal connected with Hopkin Morgan's bakery private short canal and wharf, where there was a regular procession of flour delivery boats. The next place reached is the wharf outside the Queen's Hotel, where William R. Davies had his canal stores. He and his son John had their own boats and carried cargoes daily between Pontypridd and Cardiff. On the other side of the canal was the high building of the Corn Stores, established in 1850 by William Lewis, which traded for nearly a century. Supplies were winched up from the boats, and corn was ground by steam-powered engines. The Queen's Hotel was one of the places where people would spend time watching the boats entering and leaving Trallwn Lock, No.30. (The canal men didn't use the Queen's though, the Llanover Arms next door was their particular haunt.) The spectators would also watch the loading and unloading along the wharf.

Upon leaving Trallwn (sometimes called Trallwng), the canal curves to the east, passing Canal Place from where Thomas Thomas and his wife Martha carried goods to Cardiff in their private boats for several decades. The canal runs along the back gardens of Ynysangharad Road and the chain works of Messrs Brown Lenox where barges would unload coal supplies to the works. Here, about 80 yards above the upper gates of a pair of locks (Nos 31 and 32), water entered a top basin, off the canal, and was channelled into a narrowing volume for the turbines. These in turn would drive the shafts coupled to chain-making machines, before passing under the works to a large bottom basin. Wrought-iron chain cable and anchors forged by the skilled blacksmiths employed by Brown Lenox were lowered through the floor of the test house after the examination of each link, into a waiting canal boat, for the trip to Cardiff. In 1824, nearly 1,000 tons of iron products were carried on the canal. By 1835 the annual shipments had increased to 2,500 tons, and in 1839 reached 4,000 tons.

While at the Ynysangharad Locks the boats would take on as cargo beer, ale and stout from Ynysangharad Brewery and Bottling Stores at the busy Bunch of Grapes Inn near the chain works bridge. Moving down in a more southerly direction, stone from the John Gibbon Quarry behind the Farmer's Arms Inn was to be sent from this point.

Rails made at the Taff Vale Ironworks just over the river were sent by drams across the 1851 bridge for loading on to canal boats near the Llanbradach Arms Inn.

Vegetables, fruit, bread, corn, rugs, carpets, furniture, pianos and clothes were all handled along this stretch with Trallwn at its centre, and this is why Pontypridd displaced Llantrisant as the market town of mid-Glamorgan.

There were 200 boats on the canal in 1830, some owned by the GCC and some were owned by local carriers. The tolls paid by the carriers were calculated on their written declarations of the cargo, endorsed by the toll collector. He could check the tonnage by gauging the depth at which every new boat floated when empty, and then marking figures at several levels on the boat with notches or iron plates once the depths were determined under various tonnages. Later the figures were recorded in toll books.

At Tongwynlais Lock the GCC erected a weighing machine made by Brown Lenox which simplified checks for overweight cargo. This machine was moved to Crockherbtown Lock at Cardiff, then over to Kingsway Lock where it stayed until the end of the canal's life. It has since been moved to Stoke Bruerne Canal Museum in Northamptonshire.

For many years the tolls were 1*d*, and then 2*d*, per ton per mile for coal, iron, stone, brick, tile, sand, gravel and manure, until the coming of the railways weakened the company's bargaining position.

The GCC had its own warehouse at Trallwn, along with an extensive canal-side community, and later, as the advent of steam coal and the growth of Rhondda exports all around the world in the 1840s brought more work to the area, a replacement for the old bridge over the Taff built by Mr Edwards was needed. This obsolete bridge had rapidly become a hindrance to traffic on the canal, and was bypassed by a new bridge, built with public funding. To everyone's delight it was built alongside the old one, which meant that no roads had to be altered, which would have caused more expense.

Moving on to the 1870s at Trallwn, Thomas Thomas was still operating, and J.J. Thomas & Davies & Son had a regular daily timetable as canal carriers between Trallwn and Cardiff. Six horses were stabled here for the GCC deliveries to town and district. When the Marquess of Bute took over the canal in the mid-1880s, the company commenced carrying to Abercynon and Pontypridd, and at this time the previously mentioned Martha Thomas had become the agent for the GCC at the Trallwn canal office.

Several pubs, hotels and beer houses sprang up about the canal here, including the Llanover Arms, Crown Inn and the grander Queen's Hotel, and it was to these wharves that the steam boats of the canal company were introduced from Cardiff to Pontypridd. Unfortunately, they were unsuccessful: they needed more depth of water, the canal bridges were too low and there were too many locks to negotiate to make a quick journey.

HOPKIN MORGAN AT THE EAST STREET BAKERY

Hopkin Morgan was born in 1854 on the Graig, where his father kept a grocery shop in the High Street. Bread was baked in the shop by his mother, but later a bakery was built further up the hill in a building which is now a club. Hopkin Morgan then opened a large steam-powered bakery in East Street, Trallwn, where a basin and a connection to the Glamorganshire Canal was made so that boats could unload their cargoes of flour in sacks brought up from Cardiff. Hopkin Morgan opened shops or delivered to other shops all over Glamorgan. He was well known in Cardiff, and was a generous benefactor, eventually becoming a Justice of the Peace. My personal memories of this company include the building of a large garage on the corner of North Road and St Athans Road (Western Avenue) in the late 1930s to give support to his large fleet of lorries and vans. This was at the end of the canal's useful life, and his lorries were picking up the sacks of flour from Spillers, at Roath Dock.

THE BROWN LENOX CO. AT YNYSANGHARAD

The Newbridge Works were opened and operated by Captain Samuel Brown to produce chain cable for the Royal Navy initially, but was to go on to make chains and anchors for the merchant navy, and the navies of the world. Today, over 190 years later, these works are still in production. Although no longer a chain works, it is the oldest extant works in Wales. Today, the main activity of the works is the manufacture of ore- and waste-processing machinery, such as jaw and gyrating crushers, hammer mills and domestic shredding machines. A lifting gear division and testing laboratories also form part of today's engineering complex.

In 1968 the company became part of F.H. Lloyd Group, and chain cable production was sent to the group's Midland Works, where the group concentrated all their chain work, large and small. The last major chain cable order fulfilled at the Newbridge Works was the cable for the *QE2*.

Sam Brown (later Captain Sir Samuel Brown) was born in 1776 in London. At the age of nineteen he joined the navy as an able-bodied seaman on HMS *Assistance*, and worked his way up to acting lieutenant, serving with distinction during the Napoleonic Wars. Upon his return to peace he turned his thoughts to what he had seen on board ship and what he could do to make things more efficient in the Royal Navy, taking mooring and rigging as a start point. After spending several months shut away in two rooms in a house in Dove Court, London, he came up with a chain cable of wrought-iron twisted links, and he engaged a blacksmith to make the chain. In need of capital, he turned to his cousin Samuel Lenox for financial aid. Lenox was a successful merchant and he provided a stable basis for future development.

Merchant and Royal Navy vessels of the day used hempen cable, which for a ship like the *Victory* could be 20in in diameter. The disadvantages of hempen cable were many, but its replacement had to wait until wrought iron was produced in the course of the recent industrial revolution in sufficient quantity, as well as quality. Hempen cable had a short life due to the fact that it rotted away through alternate wetting and drying, and also to the cutting action of rocks and ice. It also took up valuable space on board ship, having to be stored in cable tiers by the most able men on board, so that it could be run out quickly without jamming. Another disadvantage was that it could be shot away in action, and was a health hazard by providing an ideal medium for fungal and bacterial growth.

Brown and Lenox formed a partnership in 1808, trading under the name Sam Brown & Co., and Brown proposed the use of his cable to the navy; as a practical demonstration he chartered the *Penelope*, a 400-ton sailing vessel, and fitted her out with chain cable for both rigging and mooring. Brown then captained the *Penelope* on a four-month voyage to Martinique and Guadeloupe in the West Indies.

The voyage was a complete success, as regards proving the merits of his chain cable, and a favourable report submitted to a naval committee induced the government to order the re-equipping of four warships with iron chain cable.

By 1811 chain cables were in general use on HM ships, with one cable to each ship. However, the full complement of hempen cables was still carried, and it took many years before they banished them in spite of the fact that it brought safety and an increase of 50 per cent to the locker space for cables.

Brown was promoted to master and commander in recognition of his work, and the patents he took out gave him exclusive rights as sole supplier to the Royal Navy.

Brown now looked for a place to build a factory, and Newbridge (Pontypridd) was his final choice as he had been happy with the iron he had been using, and that came from Crawshays Ironworks at Merthyr. Iron brought down from Merthyr by canal boat was secure, and the local knowledge of his foreman of smiths, Phillip Thomas, confirmed his choice. Phillip Thomas had been involved with the development of chain cable design, and his contribution led to a joint patent with Brown in 1816. He later became the first manager of the works, doing this until his death in 1840.

Newbridge Works, in addition to its chain cable production, also made chain for many suspension bridges, and Brown's innovation was that of chain with flat links and pins. A design he patented in 1817. Telford was to use this design later on the Menai and Conway bridges.

The first suspension bridge erected by Brown still stands today – the Union Suspension Bridge over the Tweed, near Berwick-upon-Tweed. In all, Brown was associated with around forty suspension bridges and piers, the most famous of the latter being the Brighton chain pier that opened in November 1823. Only two suspension bridges of Brown's design have been recorded in Wales: one at Llandovery over the Towy and another over the Usk at Kemeys – both now replaced.

In the field of ships' cables, the company built up its reputation as the premier manufacturers of chain cable. One of their most famous orders was for chain cable for Brunel's ill-fated ship the *Great Eastern*. Brunel visited the works several times.

The company made buoys, anchors, swivels and forgings for many uses, and by this time were using iron from Anthony Hill's Plymouth Works, but later started to make its own iron, making rolling mills and foundries for cast iron, crucible steel and brass.

At the turn of the century the works were making chain for battleships with $4\frac{1}{2}$in square links, requiring hydraulic presses for shaping them. Other famous ships with chain cables made at Brown Lenox included the Cunarders *Mauritania* and *Aquitania*, and the battleships HMS *Lion*, *Dreadnought*, *Hood* and *Rodney*.

The Newbridge Works continued with this work of high-quality chain cable and anchors well into the twentieth century, and turned out various engineering products such as colliery winding engines and a variety of castings and mouldings. There was a flourishing trade in iron buoys made on contract to Trinity House Corp. In 1922 they installed a foundry to handle cast steel, where they set out to produce continuous cast-steel cable, and in 1928 received approval for this cast-steel cable from Lloyds.

In 1958 they installed the largest, most modern semi-automatic chain-making plant in the world. With this development wrought-iron chain ceased to be made at Pontypridd after 150 years of continuous production. In 1968 the chain-making resources were moved to the group's Midlands factories. Appropriately, in the tradition of a company that had supplied the most famous ships in the world, Brown Lenox's last job was to make the anchor cables for the *QE2*, before closing down its chain-making era.

In 1969 Brown Lenox became a wholly owned part of the F.H. Lloyd Group, and with it came a change of products.[1]

THE CYFARTHFA CANAL

There was an earlier small canal at Cyfarthfa which was cut in the mid-1770s by Anthony Bacon, something like twenty years before the Glamorganshire Canal was proposed.

It ran from a junction with the Canaid brook near the later Cwm pit, to the Cyfarthfa Yard. Charles Wilkins tells us that:

It was so arranged that each of the coal levels, or coal holes were cut on the same contour line so that they could be serviced by this canal, with a small bay or dock in front of each. The water level was maintained carefully so that water did not flood the level and workings.

The boat was handled by two men or women, mostly in a string of several boats, one person on the bank pulling with the aid of a harness over the shoulder, and the other in the first barge, with a large boathook to either push the string away from the bank or to pull the barge to shore.

The barges were made of iron, and they shipped as much coal as had been deposited outside each bay. A hard day's work was guaranteed. This canal ran roughly parallel to what would be the line of the Glamorganshire Canal, which would arrive later, and it fell out of use sometime between 1835 and 1840.

This tub-boat canal was no doubt supplying Cyfarthfa Ironworks, as the hillside west of the canal was where the works obtained their ironstone, and the coal was to go to the company's coking yards, and furnace charging platforms at Cyfarthfa.

This canal wound its way south along the 650ft contour as far as Cwm Canaid.

ENDNOTE

1 Stephen K. Jones, 'The History of the Newbridge Works of Brown Lenox & Co.' in Stewart Williams, *Glamorgan Historian*, Vol.12.

When the northern stretch of the canal closed and water levels decreased, the whole cut returned to nature and the towpath became a country walk. Looking north, this, the first of three bridges, was the Vale of Neath railway bridge, engineered by Brunel. The railway ran from Merthyr, through Aberdare, to Swansea Docks. The next bridge, further north, is of the Brecon & Merthyr of 1866. The third bridge, in the far distance, is the cast-iron bridge that carried the parish road across the canal in front of Rhyd-y-Car Farm. This is the bridge that was preserved and placed in the museum site. (Merthyr Library)

Brunel's Vale of Neath Bridge today. The other two bridges, and all traces of the canal bed, have been bulldozed as the whole area is part of a large business park at Rhyd-y-Car. (I. Jones)

This oblique RAF photograph, numbered 16050, of 7 March 1945, shows the short length of the Glamorganshire Canal between the railway bridge of the Vale of Neath line which can be seen in operation curving north to Merthyr Tydfil station. The River Taff keeps company with the canal a little to the north-east, with Brandy Bridge over it to the right. This was the site of the mighty Plymouth Works. The Pentre Bach and Duffryn works were a little further south. The remains of the Dowlais Iron and Steel Works are at the top centre and the Ivor Works are to the north. The Penydarren Works were sited at the top left corner of the photograph. This was a weird world of cinder tips, slag heaps and spoil dumps, railways and tramroad inclines.

The canal can also be followed further to Glyndyrus Lock at the bottom middle, and continues to the left of the Vale of Neath Bridge to pass under the Brecon & Merthyr Bridge and the Rhyd-y-Car Bridge before heading north.

In the middle of this photograph, running from the north bank of the River Taff, up to near the top of the photograph, is the Dowlais Bridge Railway, which took spoil and cinders to the tips. (Welsh Assembly Government Photographic Archive)

Glyndyrus Locks Nos 4 and 5. This photograph was taken after the upper section of the canal had closed, c.1900. Beyond the stable and lock house on the left was the GCC's Glyndyrus Reservoir. This is the southern end of the lock and all has now gone, filled in and partially levelled, but one can still see the rise in the towpath which is now a cycle path. The real story of these locks is the way the buildings have been utilised. The lock house is now a fine, modern, remodelled dwelling which, along with the stable and rest house, look well. There is also the boathouse which looks like an arched tunnel that goes back to the beginnings of the GCC. It is long enough to accommodate a canal boat and gains access to the canal at the top lock level. It must have come from the canal into the lock house porch at an angle. (Merthyr Library)

The entrance to the boathouse is the arch, filled in by the present owner with his own masonry and a door, but inside is the tunnel-shaped boathouse, now used as a storeroom. (I. Jones)

This is the inside of the Glyndyrus boathouse. Access must have been at the high part of the rise of the lock – between locks 4 and 5. As can be seen, steps to the road level have been introduced since the canal's closure. Why was the boathouse here? I do not know, but, it could be that the lock keeper was a 'lengthsman' responsible for a considerable length of the canal's maintenance. (I. Jones)

Moving south from Glyndyrus Locks through lovely valley views, we come to Upper Abercanaid, a coal community. Fronting the canal is Quay Row, a terrace that was a busy wharf at one time, loading coal on to canal boats from the famous Waun Wyllt Colliery and several other coal levels around this area. The Plymouth Ironworks had a canal basin here that also collected coal for its works across the valley. (I. Jones)

This is what is left of the canal at Upper Abercanaid, a shrunken watercourse maintained by a spring. (I. Jones)

One of the bridge abutments on the west side of the canal, which carried a tramroad over the canal from coal levels to the basin at Upper Abercanaid, and later became part of the railway system. This bridge was probably 'Pont Racks', just a short way from the village. (I. Jones)

Moving another mile further south, we are at Abercanaid, another mining village. RAF 58/676 gives us an overhead look at this once-important canal-side village. The Glamorganshire Canal runs across the middle of this photograph from left to right. The River Taff runs across the top. Gethin No.1 pit site is at the top right, with Gethin No.2 at the bottom right, and the old track of the incline shown joining the two. The site of Graig pit is at the bottom left and the cross in white marks the spot where the Llwynyreos Inn still operates. (Welsh Assembly Government Photographic Archive)

This photo was taken from the middle of the Pwllhwyad Bridge in Rhydyfelin looking north towards Pontypridd. The picture must have been taken prior to its demolition in the 1950s.

3

The Coal Champions

The construction of the Glamorganshire Canal was sponsored by the ironmasters to move their product to a port for sale. It was not envisaged that eventually coal would become the more important user of the canal. Due to the energy and persistence of a few men of vision in the area of the Taff Valley, Rhondda and Cynon, the canal had its hands full, even when the railways came.

In the early years of the canal coal was only being used by the ironworks for its smelting, coking and smithing. Small amounts were used by the local population in the towns where it was delivered by horse and cart from the levels near them, and if the ironworks had a surplus of coal over their immediate needs then it was sent down to the towns along the canal for sale, but it was not a large business because of its priority as conversion to coke for iron smelting.

In the period from 1700 to 1790 a small amount of coal was being mined, and exported from Cardiff at the wharf on the River Taff at Quay Street coming from levels in the lower Taff Valley. The ports of Swansea and Newport were shipping much more, but this was to change as the years went by. The change in exports from Cardiff was slow but sure, and Cardiff was not only going to surpass these rivals, but become the biggest exporter of coal in the world. For this we can thank a few entrepreneurs and developers who were to inspire others.

First among these was Mr Walter Coffin (1785–1867), who was to initiate the idea of 'sale' coal (coal specifically for sale, not connected to ironworks) in the area of Cardiff, with the sole intention of substantial exports of coal from that town. He was born in Bridgend and educated at Cowbridge Grammar School, followed by a private academy at Exeter. In 1804 he returned to Bridgend to help in the family tanning business. Unsurprisingly, he did not like this work, and he began to consider a career selling Welsh coal. His father – also named Walter Coffin – had invested in land, purchasing farms in the parish of Llantrisant.

In 1801 he bought from William Humphries the Dinas Uchaf Farm in the north-west corner of the parish, and eight years later, in 1809, Walter Coffin the younger gave notice to Lewis Robert Richard to terminate his tenancy of Dinas Uchaf Farm, and with financial support from his father prospected for coal on this estate in the lower Rhondda.

His first level to the Graig Vein near the farm was not successful, for the seam was thin and of inferior quality. His second level, opened in 1809 to the Old Vein of No.2 Rhondda, located a seam of good quality, about 3ft in thickness, which prompted Coffin to extend his mineral base and sink a vertical shaft. It was at this Dinas Lower Colliery, opened in 1812, that the Bodringallt Vein (No.3 Rhondda) was found at a depth of 40 yards. This was the seam that was to make Coffin's fortune, a seam of the finest bituminous coal in South Wales. Coffin's next task was to get his coal to Cardiff's seaboard.

The Glamorganshire was the obvious route, and in 1809 one of the earliest members of the canal committee, Dr Richard Griffiths, had leased part of his Hafod estate to Jeremiah Homfray to open a coal level there. The doctor then built his famous dram road for 2 miles and, after bridging the Taff, gained access to the canal at Dynea by building his own canal from the bridge to the junction with the Glamorganshire Canal, completed in September 1809. Within three months of this, Water Coffin was meeting the commissioners of the canal to discuss rental terms to be paid to neighbouring landlords in his project for a dram road to link his Dinas Levels to the doctor's dram road. The link was achieved by the end of 1810. These two men entered into an agreement which ensured that virtually all coal raised in the lower Rhondda would have to be carried over their interconnecting lines.

Coffin then sought markets for his coal, by conducting personal interviews with prospective purchasers in southern Ireland and elsewhere, as he acted as his own agent of a small company. He secured a coal wharf at Cardiff ship canal and also bought some canal boats, and then a brig named *Brothers* of 79 tons, the first of his coastal vessels for use in the coal trade, cutting out the middleman – the ship's master – from his part of the coal trade. By 1820 Coffin was the second-largest shipper of coal on the Glamorganshire Canal, exporting 10,564 tons, a figure only exceeded by J. Bennet Grover with 15,481 tons from his Maesmawr Colliery, near Llantwit Fardre.

In 1828 the Cardiff Harbour report gives the coal shippers' tonnages as:

Walter Coffin	23,662 tons
Charles Smith	10,660 tons
Revd George Thomas	8,888 tons
B. Grover	7,585 tons
C. James	2,447 tons
Morgan Thomas	2,200 tons
Mr Llewellin	1,175 tons
Total	56,617 tons

In the 1830s, several new rival shippers/producers of bituminous coal had appeared on the canal wharves at Cardiff, notably Thomas Powell and George Insole. Powell, who already had a small mine at Aberbeeg, sank two shafts to a 6ft vein at Gelligaer. He connected his colliery to the Glamorganshire Canal by a tramroad, and in March 1830 leased a 318ft wharf at Cardiff from the Marquess of Bute. Next, he built a jetty at the western end of his canal holding on to the River Taff, and this enabled him to load coal on to vessels directly when tides were favourable. Powell was a formidable rival, for by the mid-1830s, some years before he became the biggest exporter of steam coal, he was beating Coffin into second place for the export of bituminous coal in the port of Cardiff. There is also evidence of George Insole canvassing for orders in southern Ireland in 1833 – a market that Coffin had long regarded as his own.

As the port of Cardiff became busy, exports of iron, imports of iron ore (the ironworks at Merthyr were not only running short of local ore, but they were finding that the imported ore was better) and the export of coal led to congestion and delays. So frustrated was Coffin that in June 1835 he left the canal committee, and in October he became a member of the provisional committee formed to inaugurate the Taff Vale Railway.

Although the main objective of the TVR was to carry iron from Merthyr to Cardiff, the TVR Act of 1836 incorporated a clause for the construction of a branch railway to Coffin's pits at Dinas. The railway opened from Merthyr to Cardiff in April 1841, with a branch line from Newbridge (Pontypridd) to Eirw in the lower Rhondda. In June 1841 the Dinas tramroad was linked to the TVR at Eirw, and the direct link from Dinas to Cardiff was forged.

THE INSOLES

George Insole was born in Worcester in 1790, the son of cabinet-maker William Insole, and George probably served his apprenticeship under his father in that trade.

Probably because of the well-published accounts of the Klondike at Merthyr, George, who was now a family man, decided to move to Cardiff with his wife and children in 1827, where he formed a partnership with Richard Biddle in a business selling timber, coal and bricks from a yard in Cardiff. Sales were reasonably successful until three years later when the partnership was dissolved as Biddle was declared bankrupt from previous dealings. Insole reckoned that he had been swindled too by Biddle. Henceforward, Insole was to concentrate on the sale of coal; by February 1830 he had acquired a yard and wharf at the southern end of the sea lock pound to set up as a coal shipper.

As the large collieries that existed then were under the control of the ironmasters, the smaller levels and pits were supplying the house coal and the coastal trade in the Severn estuary, around the west coast of England and Wales, and to Ireland with bituminous coal. The newer trade in steam-raising coal which extended to London was at that time being supplied by the semi-anthracite coals of West Glamorgan, and being shipped from Llanelli and Swansea. The coal trade at Cardiff and Newport was based on its bituminous coals which were used not only for house heating and cooking but for blacksmiths, commercial undertakings and industry.

George Insole set out to sell in all of the above areas, plus the growing market for steam coal. His bituminous coal came from several collieries in the Taff Valley, the most important of which were the John Davies Gelliwion Level near Newbridge (Pontypridd) and David Morley's Craig-yr-Allt Level near Taffs Well.

In 1830 Insole started to market Waun Wyllt steam coal from Robert Thomas of Abercanaid. This was excellent steam coal and the pit sold all it could mine. Robert Thomas died in 1833 at the age of fifty-eight. His widow Lucy Thomas continued his business with her son William, gaining a unique place in the history of the coal fields. Neither Lucy nor Robert could write but they proved to be shrewd and successful.

The first two boatloads that Insole ordered for his first venture into steam coal of Waun Wyllt sold readily in Ireland and south-west England, and Insole realised that the qualities of this Merthyr 4ft seam was a winner. With this in mind he attempted to enter the London market, and provide the Thames tugs and packet boats with suitable steam-raising coal, and he was receiving a boatload a day from Lucy Thomas.

He had sent a cargo of steam coal to Malta in 1831, and in the same year bunkered HMS *St Pierre*, becoming the first South Wales coal owner to supply the Royal Navy. Within a few years Insole was sending cargoes of steam coal to new markets, including Brighton, Ramsgate and, at a greater distance, Quebec and Alexandria.

George Insole had not forgotten that it was the bituminous coal that had given him his base and financial security, and he struggled to keep up with demand as the population, and thus the need for domestic coal, expanded in South Wales. On 12 October 1832 he made enquiries of Dr John Thomas Casberd of Penmark, who had inherited the property of Maesmawr, asking of the possibility of leasing the old colliery on the west bank of the Taff. This pit had fallen on hard times and had ceased production in 1831 because of the inadequacies of the previous owner, John Bennet Grover.

George Insole and Dr Casberd agreed terms, and November 1832 saw the opening of the Maesmawr Level. He also wanted to open a level on the neighbouring land of Maesbach, which was the property of the Marquess of Bute, and was not as promising as Maesmawr, but as it was near Insole perhaps wanted it for drainage and ventilation.

The marquess and Insole agreed matters such as royalties and Insole had both levels, and also found £3,000 to sink a new pit at Maesmawr, very close to the Taff. There is no evidence of coal being mined at Maesbach, but Maesmawr provided increasing amounts of coal throughout the 1830s. In 1833, output was 12,943 tons, which by 1839 had nearly doubled to 23,444 tons. George continued to build up his markets by direct selling to customers, both in his established West Country markets, including Minehead, Bideford, Truro, Penzance and Plymouth, and in Ireland which he visited every year. He also sought house coal sales to Bristol and London.

The expense that he had incurred at Maesmawr and Maesbach proved very timely, as from the 1830s a steady expansion of the iron and coal trades in the Glamorgan valleys came about. New ironworks were opened, and new coal seams were discovered; it became clear that the canal could not cope alone with a capacity of 1.5 million tons per annum.

Look out! The Bute West Dock and the Taff Vale Railway were on their way.

◆ ◆ ◆

Meanwhile Walter Coffin was having problems at Dinas in 1832. The miners had worked the Dinas Lower Colliery No.3 Rhondda seam northwards until meeting a major fault running right across Coffin's Craig Ddu property. This fault was known as the Cymmer (Dinas) Fault, which lowered the seam to the west by about 40 yards. In order to recover the seam lying to the north and west of the fault on the 20 acres still unworked, Coffin sank a new shaft, the Dinas Middle Colliery, about 600 yards up the valley. This proved to be costly, and after an outlay of £1,000 Coffin felt prompted to look elsewhere. In February 1832 he renewed his lease of 507 acres of the adjacent mineral property to Dinas at Brithweunydd, and from 1833 the Rhondda No.3 seam was worked under this property from the existing Dinas Middle Colliery, and from a new level that opened in 1839.

Despite the efforts of producer/shippers such as Powell, Coffin and Insole, the growth of Cardiff's exports was slow, still lagging behind Newport and the ports of west Wales. From 1840, however, the potential of Cardiff was quickly realised, and the day when Cardiff would be the premier coal-shipping port was fast approaching. In 1850 the increase was dramatic, reaching the total of 731,329 tons coastal and foreign sales. Notably it was due to the fact that the foreign market reached 250,000 tons.

The Aberdare Valley was the new focus of coal production, and steam coal became so much in demand worldwide that a number of new collieries were opened between 1842 and 1850 in the valley. They were not small levels but large, deep pits. The steam-raising coal from here was a most desirable commodity for railways and steamship lines.

Records show that of the 750,000 tons exported from Cardiff in 1850, half a million came from the Aberdare Valley, and most of it went to the French market.

Walter Coffin had settled into a comfortable marketing system, where his contract to supply coke made from Dinas, or Coffin's No.3 vein, to the Great Western Railway secured an important internal market from 1841. He also tried to gain new markets overseas for Rhondda coal, especially in France.

It was John Nixon who created the trade in steam coal between Cardiff and France when, in 1842, he took his first cargo of 4ft seam coal from Thomas Powell's 'Old Dyffryn' pit in the Aberdare Valley to the sugar refineries of Nantes.

Coffin had moved his wharf to the new Bute West Dock from his place on the Glamorganshire Canal, near to that of Thomas Powell, and actually adjoining that of George Insole, as a result of which he could easily gain information on the latest trends in overseas shipping.

Coffin kept to his bituminous coal deliveries, especially his No.3 Rhondda, the quality of which was assured. In 1845 shipments were directed to his two well-established markets: the coastal towns of the Bristol Channel (above all Bristol) and the ports of southern Ireland, such as Waterford and Cork. The records show no shipments abroad by Coffin until 5 August 1848, when he sent 116 tons to Nantes. This was a precursor of a flourishing trade, because before the end of that year Coffin was dispatching regularly some five to eight ships a week from Cardiff to Nantes. The cargoes were small – between 100 and 150 tons – whereas the usual size of Powell and Nixon's was 400 tons, but the work was regular.

Coffin played no part in the steam coal market. He was the owner of three flourishing collieries, the Dinas Middle, the Brithweunydd and the Gellifaelog. He had made large profits for several years. He was also the owner of considerable property in the lower Rhondda, a substantial shareholder in the Taff Vale Railway, and owned several collier ships, so in 1853 he decided to retire from active commercial life to take an interest in public affairs. His shipping interests at Bute Street were transferred to Richard Parry, while William Ogle Hunt took over his mining interests in the lower Rhondda.

◆ ◆ ◆

George Insole knew by now that he had to gain access to more coal in a major way, so he and his son James Harvey Insole leased mineral rights of 375 acres of Cymmer land, adjacent to the Dinas property of Walter Coffin. Immediately, the south Cymmer Level was opened to the No.2 Rhondda seam, and in 1847 the No.1 Pit, or Cymmer Old, was sunk down to the No.3 Rhondda seam. Just in time! The Maesmawr Levels and pit were running out of product, and Insole was struggling to meet his orders; only 1,499 tons were exported in June 1847, so the production of the No.3 seam at Cymmer was intensely worked, and ever-increasing quantities were sent daily to Cardiff.

The snag then was transport. The Rhondda branch of the TVR terminated at Eirw, 900 yards below Cymmer, and Insole's neighbour, Walter Coffin, refused to allow him the use of his private tramroad. Consequently Insole had to build his own tramroad from Cymmer to the 'Old Mill'. A wooden bridge was built over the mill brook, and coal loaded into carts was taken over the circuitous route to the TVR at Eirw. Insole kept up a long-term fight with the railway company to build an extension to their line, and in 1846 it was done; by the December of the following year production soared to 23,656 tons. The Cymmer gamble had paid off, and soon the mineral property under lease covered 1,300 acres worked by two additional shafts: the Upper Cymmer (1851) and the new Cymmer (1855). Three varieties of coal were worked: Low Main No.9 Rhondda for smelting and forging; No.3 Rhondda for coking coals; and a coal classified as 'Through and Through', most of which was shipped.

We have already mentioned back in 1829 George Insole's purchase of Waun Wyllt coal, or Merthyr 4ft-seam steam coal, and how much in demand that coal was to the Royal Navy and Merchant Marine. This led to the opening of new, large steam-coal collieries in the Aberdare Valley and ultimately to the acquisition of new overseas markets by two of the successful coal owners, Thomas Powell and John Nixon. George Insole was aware of the increasing demand for steam coal, and that much more could be sold if available. Insole had no steam-coal resources of his own, and after his recent purchases at Cymmer to prolong his holdings of bituminous coal, he had little financial resource to invest in steam coal. He instead reverted to his original trade of agent shipper at Cardiff Docks for Aberaman Merthyr Steam Coal, mined by David Williams in the Aberdare Valley, and it was again commended by the Admiralty as being the best of coals.

The consequences of these Admiralty reports on the growth of the company of George Insole & Son became clear between 1848 and 1851. Sales of steam coal were dramatically increased by large Admiralty orders, and by the developing markets in the mercantile marine. Insole was supplying the first steam-coal packet companies in the world, including the Royal Mail and Peninsular and Orient (P&O); this brought massive publicity to Insole's drive for new markets for his bituminous coal, as well as steam. Up to 1847 Insole's markets were coastal and Irish, subsequently they became foreign and worldwide.

George Insole died suddenly on New Year's Day 1851. He had laid the foundation of this trade for his sons to follow. The period of beginnings had ended; the period of consolidation was to last a long time.

Because of the provision of the Glamorganshire Canal that was built for the ironmasters, to most of the pioneers of the coal trade it was a ready-made blessing, competing with the railways for a long time.[1]

ENDNOTE

1   Some detail of this section from: E.D. Lewis, *Pioneers of the Cardiff Coal Trade*, and Richard Watson, *Rhondda Coal Cardiff Gold*.

Travel south from Abercanaid for less than a mile and you come to the northern end of the village of Troedyrhiw at Castle Houses, built on the towpath of the canal. Very old but well looked after, they housed Castle Colliery workers, the railway being higher up the mountainside. (Rhondda-Cynon-Taff Libraries)

Yet anther postcard view! Castle Colliery is on the other side of the valley to the cameraman and the canal runs across the middle of the picture. The houses at the left middle and the Castle Houses at the right show the line of the canal and the slope from the canal level to the colliery is evident. This was a Crawshay pit and the coal was taken to Cyfarthfa by the Crawshay railway. Nothing was transported by the canal. (Rhondda-Cynon-Taff Libraries)

A fine image from Bob Morrison's collection, showing Castle Colliery above the dried-up canal, ten years after closure of this upper section. The Dynevor Arms is at the left and the Castle Houses are at the right on the towpath. (Bob Morrison)

The canal at Troedyrhiw, with its paved towpath, has its new steel footbridge. This is now used extensively by cyclists as well as pedestrians from the large housing estates that have been built up the mountainside. Lovely views now exist as industries have left, as seen here between the Dynevor Arms and St John's Church. (I. Jones)

This very old stone bridge is the farm road over the canal at Pont Nantymaen on the northern outskirts of Aberfan. (Pontypridd Library)

Aberfan Locks, c.1890. At the upper right one can see the canal boat horse completing his climb to the upper towpath. The lock house on the left and the lock-tail bridge are typical of their day. It is an impressive photograph, perhaps due to the close proximity of the house to the mighty lock. (G.H. Bedford)

This is all that is left of locks 6 and 7. The house has been demolished along with most of the chambers of the locks, but with a bit of cutting the masonry was exposed. It was too tough for the contractors to demolish further. (I. Jones)

4

Canal Customers of the Glamorganshire Canal (North)

We have already reviewed the first customers of the canal, the huge concentration of iron producers at Merthyr, and their combined need of the canal. At least for a while, all the ironworks used the canal until arguments precipitated the Merthyr Penydarren tramroad to Abercynon, which some then utilised before rejoining the canal there.

As well as the iron trade at Merthyr, there were many public carriers by canal boat, offering service down to Cardiff or any stop in between. These were James, Thomas and Lewis Williams every day, and Pride & Co. and Daniel Llewellin weekly. From Jackson's Bridge, Thomas Williams and Woodman & Co. would ship loads three times a week. Goods of all kinds came up by boat from Cardiff as Merthyr, expanding each month, needed foodstuffs for people and animals. Hardware, furniture, fabrics, grain, vegetables, flour and tools were delivered to Merthyr by boat.

Then Thomas Key's colliery at Abercanaid started to sell coal by canal. He was granted a twenty-one-year lease of a small wharf (40 yards square) below lock No.2. From this yard he sold his coal to the village of Merthyr in October 1793. Some coal was sent from Abercanaid to Cardiff, where Key was selling his coal from the small levels in the southern Taff Valley.

In 1800 both Dowlais and Cyfarthfa works, which used coal from their own collieries in their furnaces, realised that any spare coal that was available could be used for sale. They both had wharves at Cardiff Sea Lock. In 1824, Robert Thomas took on the Waun Wyllt Colliery. The colliery was 320 yards from the canal so he built a tramroad under the 'four-mile clause' to a wharf next to Glyndyrus Farm. His account with the GCC was opened in July 1828 when he shipped the tramroad materials up to Merthyr. From January 1829 he was sending an occasional boatload to George Insole's yard at the sea lock pound at Cardiff.

By 26 April, Insole had received twelve 20-ton boatloads of Waun Wyllt coal. This coal was smokeless and good for steam-raising and its sale, for domestic use, was proved to be wasteful as this type of coal was much in demand. Insole sold some of this coal to London for trials on 12 November 1830. The people in London were very impressed and made a trip to South Wales to seek out the source of this smokeless coal. When they negotiated with Lucy Thomas for the total surplus output, they agreed a price of 4s per ton and were sold in London at 18d. In August 1831 Insole complained to the Thomases that they were allowing the London people to ship their coal after he had negotiated the sole rights of shipment for himself. That year it was found that a compromise was met because of a contract between Insole and Edward Wood & Co. (London) for 3,000 tons of Waun Wyllt coal.

This marketing of Insole triggered demand for coal from Merthyr. In 1830 Robert Thomas sent 10,476 tons to Cardiff and, in 1831, he applied to lease his own coal yard at Cardiff, but he continued to supply Insole. As mentioned earlier, Thomas died in February 1833 and his wife Lucy and their son William continued the business. In 1833 the Thomas's colliery sent 16,563 tons, increasing to 18,754 tons in 1834. In 1838 the widow Thomas and her son took on a further mineral lease, that of Graig Farm, adjacent to Waun Wyllt. They opened a colliery on the bank of the canal near Key's old level. There, a large basin was built for the coal boats to draw alongside the tipping wharf within the colliery confines. The mine engineer, John Nixon, wanted to get in on this Welsh steam coal at Lucy Thomas's pit, but he failed to convince. Lucy died in 1847, while the Graig pit account with the GCC continued until January 1877.

Plymouth Ironworks, needed more iron coal and wanted sale coal. They sank the Upper Abercanaid Colliery beneath the banks of the canal near Key's old level and connected it to their tramroad, which crossed the river at the ironworks. The tramroad also connected the colliery to the ironwork's own basin on the canal, immediately opposite the old boat dock of Waun Wyllt.

The Plymouth estate let the lease of Waun Wyllt to Anthony Hill and, in 1844, George Lockett contracted with Hill for 100,000 tons of Waun Wyllt steam coal at 4s per ton. From the 1870s Plymouth Ironworks had its own railway to connect to the Taff Vale Railway; nevertheless, from the 1860s the Abercanaid basin was used to ship the otherwise useless small coal from Plymouth Collieries to the Patent Fuel Works at Maindy, a few miles north of Cardiff. Such traffic continued to the 1890s.

South of Abercanaid lay Gethin Farm and in 1849, William Crawshay extended the coal supply to Cyfarthfa by sinking a pit at Gethin between the canal and the river. Gethin No.2 was later sunk from higher up the mountainside and an incline was built connecting the two collieries, passing over the canal. The Gethin collieries had their own basin alongside Gethin No.1, but it appears that when coal came down the incline, the coal drams were not emptied at the basin, but through a hatch on the tramroad over the canal where the drams were tilted and emptied into the boat below.

As well as iron and coal, there was ongoing movement of limestone to kilns and from kilns to the users. Bricks, pennant stone and later the imported ore from overseas was brought up from Cardiff, except for the cargoes brought up on the tramroad from Abercynon Wharf to Dowlais Works, Penydarren and the Plymouth Ironworks. This all ended when the Abercynon to Merthyr part of the canal closed on 7 December 1898.

A few hundred yards south of Pontygwaith lock site, we reach 'Pontydderwen' – a quiet field alongside the ruins of a building, presumably a farmhouse, on Cefn Glas land. The stone bridge is so lonely, with a road to nowhere going across! At least it is still with us. (I. Jones)

The site of another set of locks, Nos 12 and 13, under the name of Pen Locks. Again, we have no locks to view but we do have the lock house, or at least the site of it. Apparently it was demolished some years ago and a facsimile was built on the site. There are canal walls to view along from locks 10 and 11, past this house. It would appear that this lock keeper was responsible for locks 10–13. (I. Jones)

The Royal Oak public house was there when the canal was busy and remains open today. The canal flowed eastwards between the pub and the fence, until it met locks 14 and 15, where it curved southwards again, down the mighty staircase of eleven locks to the River Taff level at Abercynon. (I. Jones)

Looking inside the chamber of locks 14 and 15. It has, on occasion, been used as a rubbish dump. (I. Jones)

This photo was taken before the demolition of the lock house and most of locks 16 and 17. This space, below the locks, is the junction of the Aberdare Canal with the Glamorganshire Canal. Locks 16 and 17 are on the right and the Aberdare came in from the left. (Ian L. Wright)

Today, locks 16 and 17 are in the middle of a housing development. Indeed, these remains are in someone's garden! (I. Jones)

Abercynon's eleven-lock flight, looking up the flight to firstly, in the foreground, lock 20, the top lock of the two-rise lock 'Odyn Galch' ('Limekiln Lock'). The next one uphill is another two-rise, 'Lock-y-Waun', Nos 18 and 19, with the bridge that brings the towpath from the western (left) to the eastern side of the canal. The next pair of locks that can be seen above 'Lock-y-Waun' are locks 16 and 17. The Aberdare Canal junction is between these locks and the lower pair, 18 and 19. A pond at this junction gave access to a dry dock for boat repair. Note that a lock house was provided for each of the three pairs of locks, on the left side at each lock. (Rhondda-Cynon-Taff Libraries)

This photograph was taken from the same point as the previous photo, by Mr W. Rowlands in October 1913, this time looking downhill to Abercynon. Lock 21 is at the fore with the bridge that carried the towpath back over to the west side, and also the ancient Aberdare to Rhondda road, called Alexandra Place. The GCC stonemason yard is on the left, before 'Lock Stackhouse', Nos 22 and 23. At the very bottom of the flight is 'Lock Isaf', No.24, where a GCC boat is unloading its cargo into a warehouse. (Rhondda-Cynon-Taff Libraries)

The Glamorganshire Canal moves to the south and then to the east into this area after leaving Pontygwaith. Passing the Prince Llewelyn, a former pub, we come to the site of the Cefn Glas Lock, No.9 (1), then the canal turns south to locks 10 and 11 (2), followed shortly by locks 12 and 13 (3). Then, as the canal heads slightly to the west, we pass the canal-side public house the Royal Oak (4), then locks 14 and 15 at the top of the Abercynon flight (5). Locks 16 and 17 follow just after (6). The junction with the Aberdare takes place below the last-named locks, followed closely by 'Lock-y-Waun', locks 18 and 19 (7). Lock 'Odyn Galch', locks 20 and 21, is next (8). Further down is 'Lock Stackhouse', 22 and 23 (9). The single lock, No.24, is the last (10). Finally the aqueduct that was converted to a road bridge is at (11). (Welsh Assembly Government Photograph Archive)

The rise in the road above the car is the site of 'Lock Stackhouse', locks 22 and 23. The stonework in right foreground is part of 'Lock Isaf', lock No.24. This would have been the upper gate, the lower gate is much lower behind the camera, as this road now goes down to join the level of the aqueduct, now a bridge carrying the B4275. (I. Jones)

This aqueduct was built for the GCC so that the canal could be established on the east side of the River Taff. It had run along the Taff's west side from the canal head. This photograph was taken before the aqueduct was altered to a road bridge (B4275). (Pontypridd Library)

The aqueduct today. To the left (just out of shot) is the railway bridge that once carried the Taff Vale Railway's incline. To the other side is the site of the basin that served the tramroads and water feeders. (I. Jones)

In this photograph by Ian L. Wright, we see the Navigation Bridge and House with the portal on the right, and the dried-up canal on the left. (Rhondda-Cynon-Taff Libraries)

The line of this road, from the pavement to the fence, was that of the canal after passing under the bridge at the Navigation Inn. The Llanfabon and Merthyrs were alongside the right of the canal, on the other side of the fence. The street is named Martin's Terrace, one of the earliest streets in Abercynon. (I. Jones)

This fine ink drawing, probably by I. Phillips, shows the busy GCC wharf at Navigation (Abercynon) with the canal at the foot of the drawing – although a liberty has been taken with the design of the canal boat. The street on the right is Martin's Terrace and on the left were the company's workshops. They extended much further south out of the drawing, and included blacksmiths, carpenters etc. The drawing accurately portrays the town of Abercynon in the period of that financial wizard Mr Solomon Andrews of Cardiff.

This was the Dowlais Cardiff Colliery in 1905, situated on the west side of the canal about 300 yards south of Martin's Terrace. The canal and wharf are at the foot of the photograph. This became Abercynon Colliery by 1900, after shafts were sunk in 1889 and 1896. By the turn of the century it had been taken over by Guest, Keen and Nettlefolds, and in the peak years of the 1920s it employed 2,500 men. In 1930 Welsh Associated Collieries took control but they were themselves absorbed into the Powell Dyffryn Steam Coal Co. It was then known as Abercynon Colliery. They remained owners until nationalisation in 1947. Closure came in 1986. (Author's Collection)

From Abercynon, we move a short distance south until we meet the old bed of the canal on the side of the western embankment of the A470 trunk road. Just north of Cilfynydd on Graig Evan-Leyshon Common we have half a mile of canal bed protruding from under the A470 – indeed, large tubes of concrete come from underneath the road and enter the canal at its southern end, as if being used as a storm drain. As the canal is under the A470 for most of its length from Cardiff, it's welcome to show itself for a spell. The canal walls have survived but the canal bed has accumulated a lot of growth and leaves. Trees grow all along its length and the towpath is extant. (I. Jones)

Albion Colliery (seen here in 1900) at Cilfynydd had a massive accident in 1894 when an underground explosion resulted in the death of 287 men. This place was also of some concern to the GCC as they had suffered regularly from subsidence from as early as 1890. In 1915 a serious breach occurred beyond the footbridge in this photograph. To the right of the bridge is lock 25, Ynyscaedudwg. The village of Cilfynydd was created during the four years after the colliery opened, when over 500 houses were built. By the time of the canal's closure the Taff Railway had long been the colliery's transporter. (Curtis Postcards)

Another postcard view, this time of Coedpenmaen, just north of Pontypridd. Lock 26 is shown (Tyn-y-Graig) in 1916. The tiny lock house is at the top of the slope to the canal. (Author's Collection)

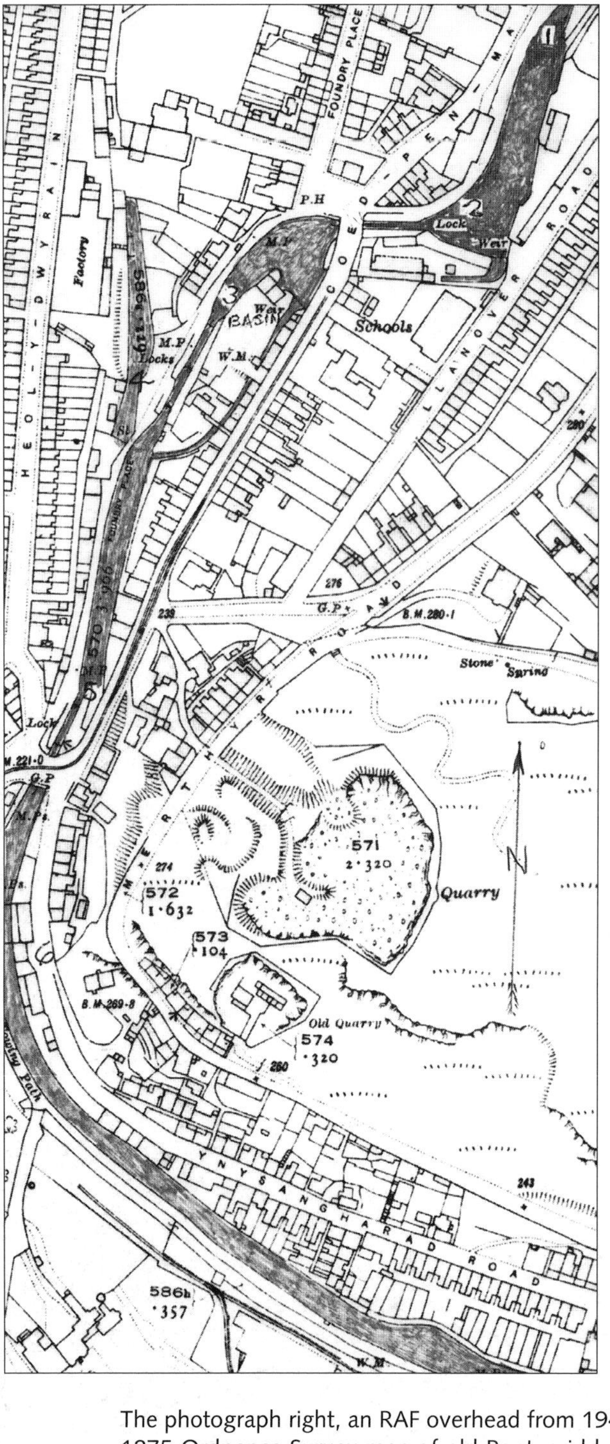

The photograph right, an RAF overhead from 1947, is best seen alongside the accompanying 1875 Ordnance Survey map of old Pontypridd. The photograph shows the filling-in of the canal, post-war, so the centre of this bustling town's streets can be organised without the locks and bridges that had been here so long. (1) Tyn-y-Graig lock 26, in the top right of the photograph; (2) Road Lock, 27; (3) Lock-yr-Odyn, 28 and 29; (4) Hopkin Morgan's basin; (5) Trallwyn Lock, 30; (6) Canal warehouses. (Photo: Welsh Assembly Government Photographic Archives)

Trallwyn Wharf in 1943. The canal company's warehouse is on the left. The canal is now a dump and is being filled unofficially. (Cardiff Central Library)

Trallwyn, Pontypridd, in 1895. This is one of two steam barges operating on the canal in an attempt to stimulate trade from here to Cardiff. This was Lord Bute's way of getting something out of his purchase of the Glamorganshire and Aberdare canals. After much money was spent on locks, bridges, wharves and canal deepening, this was his last attempt before admitting failure. The steam barges had to lower their funnels in order to pass beneath the bridges that were built long before. The Queen's Hotel is at the right and the canal company's wharf is at the left. (Cardiff Central Library)

These are locks 28 and 29 at Trallwyn, showing the depth of this two-rise construction, just before demolition and levelling. (Cardiff Central Library)

Trallwyn Bridge at lock 30, during the widening of the bridge and the strengthening that required new stone abutments so that widening could take place on the downstream side. The old stone parapets are being replaced by wrought-iron lattice work. William Lewis' 1850s Corn Stores are behind. The electric tramway from Pontypridd to the Albion Colliery at Cilfynydd used this bridge when it opened in 1905. (Cardiff Central Library)

Leaving the wharves at Trallwyn, the canal turns in a curve to the east as it heads for Ynysangharad. The water one sees in the canal on the right is not to be there much longer. This photograph shows the rear of Canal Place in 1945. (Rhondda-Cynon-Taff Libraries)

5

Canal Customers
South of Pontypridd

THE WESTERN SIDE OF THE TAFF VALE

Taff Vale Ironworks & Treforest Tinplate Works

The entire development of this village can be attributed to the industry and capital of the Crawshay family of Merthyr Tydfil. Their association with Treforest began in 1794 when William Crawshay II purchased a plot of land at Ynyspenllwch on which stood a small mill for the rolling of tinplate. This enterprise, with the aid of the Glamorganshire Canal which opened in the same year to transport its products, became an immediate success. The following years saw the expansion and modernisation of the tin works until, by 1836, it became the largest in Britain.

During this period the Crawshays had also acquired a small ironworks by purchasing adjoining land to their existing land at Treforest. They had expanded to form the Taff Vale Ironworks which, in conjunction with the tin works, formed the Fforrest Works. By 1835 these works were under the direct control of Francis Crawshay, who had moved from Hirwaun to Treforest and lived at Forest House, which is now the site of the University of Glamorgan.

William Crawshay II then opened up the Gwaun-yr-Eirw Level near Hafod to supply the two parts of the Fforrest Works and the Hirwaun Ironworks with coking coal.

With the death of William Crawshay I, William Crawshay II took over the empire of the family at Cyfarthfa and lost interest in Treforest.

The Treforest Tinplate Works originally sent its tinplate by tramroad along the west bank of the Taff to connect with the doctor's or Thomas's tramroad. Iron for the works to roll was brought down from the company's Hirwaun Ironworks by tramroad, then the Aberdare Canal, then the Glamorganshire Canal to either the wharf at Pentrbach or Pwllyhwyad Wharf. Both wharves had tramroad connections with Treforest. From the 1840s, finished tinplate was delivered from the works to a new wharf on the Doctor's Canal at Dyffryn Bridge, which was equipped with a crane for loading tinplate boxes into the boats that were to proceed down to Dynea, where the Doctor's Canal joined with the Glamorganshire Canal, and then to Cardiff.

The Treforest Tinplate Works were later to connect with the Taff Vale Railway with a siding to the south of the works. Little trade was done by canal after the wrecking of Julia Bridge, which spanned the Taff, carrying the tramroad to Dyffryn Wharf. This wooden bridge which succumbed to a storm and flood of the Taff, was later replaced by a stone bridge. The main part of the tinplate works can still be seen, just driving over the bridge. Interestingly the remains of the river bridge buttresses of the Cardiff Railway are also nearby, so it is a real industrial archaeological spot.

THE EASTERN SIDE OF THE TAFF VALE

Dynea Colliery, Pant Drain Farm on the Upper Slopes of North Mynedd Meio

From the earlier story of the coal level at Bryn Tail, we move a mere half a mile south to Dynea Colliery which occupies a steep site, where several levels have been worked in the past, leased by Walter Coffin in about 1808 for coal extraction. It was a difficult site as it is in a narrow and deep valley, scoured at its deepest by the lively stream, the Corrwg, and it is here at the upper end of this ravine that level three, which worked a 24in seam named 'Dirty Rider' that was just at stream level. Number four level is near the gate of the farm down at the bottom, which worked a seam, 'the dirty seam', of 14in of coal, 21in of dirt and 15in of coal. Another level below this still has a brick-arched entrance and an iron grid to stop the entry of animals. There are air shafts around this mountainside and the missing levels one and two suggest that there is more to see.

The present incumbent young farmer at Pant Drain doesn't know where the tramroad from the levels on his land was headed. When I suggested Dynea, he wasn't impressed but when I suggested Tyn-y-Wern he thought it could have been, but really when he looks at the maps of that period, it could only be Maes-yr-aul Wharf and Bridge where this tramroad ended. I was shown the spot where the Pontypridd, Caerphilly and Newport Railway built a bridge over the tramroad and the direction it followed pointed to Maes-yr-aul, even though it is now covered by houses.

I have made contact with two men who were young lads at the closure of the canal and the both recall the wharf at Maes-yr-aul, which was known as 'Farald Bridge', and that it was higher on the north bank and was bricked up and levelled as a tipping wall would be.

Perhaps we can assume, if not prove, that the coal was brought down on the tramroad from Pant Drain did take this course to Maes-yr-aul for shipment to Cardiff. If so, then it shared this wharf with the high explosives stores that also used Maes-yr-aul.[1]

MOVING ACROSS THE TAFF VALLEY

The West Side of the River to Maesmawr and Maesbach Farms and their Collieries

Maesmawr Farm has a history of coal mining going back to 1697 and, as time progressed the estate delivered many pits and levels. This is in the parish of Llantwit Fardre and the ground climbs steeply from the river plain up to a considerable escarpment with the farm on top. The problem here, with the coal excavated, was how to reach the nearest wharf on the Glamorganshire Canal. Coal had to be conveyed down the hill, across the valley floor for half a mile, then the trams had to cross the River Taff to the east bank. From there, they were transferred to a tramroad that took the coal to the nearest tipping wharf of the GCC.

The first serious coal mining from Maesmawr land to the wharf and bridge named 'Weaver's Bridge' at that time was by Mr Thomas Key in 1797–99, and later by his son John Key.

Maesbach Farm adjoined Maesmawr at its southern border and in April 1791 the Melingriffith Tinplate Works is recorded as having purchased coal from James Jacob of Caerphilly, the landowner at Maesbach. In 1799, Jacob allowed a twenty-one-year mineral lease to John Scott of Cardiff plus 3*d* per ton for any quantity of coal or Culm raised, exceeding 2,000 tons in the year. This was coal to go by tramroad and canal to Cardiff. By April 1805 a Mr James Morrison was working Maesbach Colliery and from the canal company's books and accounts he was finding it hard and was not financially successful as Mr John Key next door at Maesmawr.

Both of these men had wharves at the sea lock in Cardiff and Morrison was in debt to the GCC to the extent that they detained his two boats and allowed him £25 for his wharf, the sale of boats and wharf to be set against his debts. The remaining nine years of the lease were auctioned according to this description:

> Maesbach Colliery consisting of complete Iron tramroads from the two levels to the side of the canal (without need of machinery) extending one mile and a quarter. This newly opened colliery has been worked by unskilled tenants who have left the two levels injured. A recent survey by eminent miners show that the valuable coal called 'Maesmawr vein' takes its course through the whole of the Maesbach estate consisting of 114 acres and is about four feet in thickness and of good quality. There are also two other veins called the 'greater' and 'little' veins of strong quality and free burning.

It is not known whether this sales pitch made for a good purchase but the Maesbach lands appear to have passed over to the Jones family of Bristol and then, in 1808, it was inherited by a Mr Brockett Grover through marriage, who eventually combined the collieries of Maesmawr and Maesbach into one concern.

By now tramroads ran from the two estates leading to the west bank of the Taff, where two drams at a time were ferried over the river to the opposite (east) bank. The drams were hauled by mule or horse from the rail wharf to the bridge and coal wharf at Weaver's Bridge.

Brockett Grover had inherited all of Thomas Key's coal empire because he had business connections for many years with the Key family. Thomas Key had two sons and John carried on his father's work after Thomas's death. Brockett Grover married John's daughter and, after John's death in 1808 at the age of sixty-two, he inherited the lands and leases. In the next couple of years he adopted the name Key and bought more property. He had already bought Pen-y-groes Colliery on the opposite side of the Taff Vale. He then bought Maesbach so, by 1813, he owned and was operating at least four collieries and was one of the foremost suppliers of coal on the canal until his own death in 1831. He had a house at Upper Boat named 'Porth-y-Glo' close to Weaver's Bridge and next to the Upper Boat Inn. From his house he could supervise shipments of Maesmawr and Maesbach coal from the wharf at the bridge and from the bridge to Pen-y-groes Colliery. From there another tramroad ran to Tynywern Bridge. It was not long before all this activity at the bridge caused the name Weaver's to be altered to Maesmawr. However, there was to be a downward trend coming.

The acquisition of most of the Maesmawr lands by the Marquess of Bute curtailed Grover/Key and the 1824 Bute estate survey gives no indication of coal mining under Maesbach. Grover's output was reduced and in April 1831 Maesmawr Colliery was put up for sale by Grover.

Now on to the scene comes a man that would become a coal owner on a large scale, George Insole. In November 1832 he purchased Maesmawr Colliery. At that time he was a mere coal merchant, starting a career. As well as working the level at Maesmawr, he sunk a pit that cost £3,000 to the 4ft seam on the river bank, between the river and the line of what is now the Taff Vale Railway and near the station at Treforest Trading Estate. He sunk a second shaft 50 yards to the north. Both of these collieries, when shut down, had new houses built on them and these houses can be seen today.

Maesmawr Colliery was situated halfway up the hillside between Maesmawr farmhouse and the Taff Vale Railway and turned out to be the downfall of Brockett Grover. There were at least two more coal levels on this land of Maesmawr as shown on the Ordnance Survey maps of 1901, and one of these is sited just above the tracks of the main line to Pontypridd from Cardiff. It is a fairly new level that can be seen today with a stone-built arch entrance. The National Coal Board, a few years ago, boarded up the entrance for safety reasons. This level had a tramroad that was unearthed at the time by the National Coal Board and joined the old Maesmawr–Maesbach tramroad to the bank of the Taff. It has a clear tramroad trench to the bottom of the old incline.

After the Marquess of Bute acquired Maesbach he had a survey of the mineral potential of his estate carried out. This resulted in the 1830 terms for a thirty-year lease of Maesbach coal at 9d per ton to Webb & Co., whose existing collieries in Monmouthshire were shipping from Newport. Webb was allowed two years to prove the coal level, which is an indication of the state of this old colliery. They also had permission to build a bridge over the Taff and to lay a tramroad to the canal so that the journey to Maesmawr and the ferry could be avoided. A clause in the lease stated that all the coal raised was to be shipped from Lord Bute's own harbour and wharves because Lord Bute was already planning to compete with the facilities of the sea lock pound of the GCC by building the Bute West Dock.

The lease was obviously rejected by Webb as the next enquiry about Maesbach coal comes from George Insōle at Maesmawr. He bargained over the deeper coal under Maesbach and offered to reopen the now defunct colliery. He offered to work 30 tons per working day and pay a royalty of 10*d* a ton for the 4ft vein and 6*d* per ton for the 3ft vein. He also planned to bring much of the Maesmawr coal to bank through the Maesbach Level for which he would pay a wayleave of 1*d* per ton. Bute and Insole failed to agree and George Insole confined himself to Maesmawr's pits and levels to make him rich, sending his coal down the tramroad built by Brockett Grover to Upper Boat ferry, then on to what was by that time Maesmawr Bridge and Wharf.

The Taff Vale Railway Act was passed in 1836 and changed the whole perspective of Maesbach as the railway would be passing through Maesbach land. Insole now seemed to have agreed with Lord Bute's terms for Maesbach coal and he was making good money at Maesmawr – by 1841 he had 157 men at work.

The account books show that George Insole was paying royalties to Lord Bute and sending coal by tramroad until after 1846 when a siding was made off the Taff Vale Railway to Insole's Maesbach Colliery, so the old tramroad to Upper Boat was running on borrowed time.

In 1865 J.W. Booker's Melengriffith Company sank Rhydhelig Colliery on Maesbach Farm but its total output was intended for sale coal and was carried by rail on the Taff Vale Railway and Great Western Railway and played no part in the canal life. Some evidence of this colliery can still be found.

In 1844 George Insole moved his interest and money to Cymmer in the Rhondda, where he joined in competition with the big players in the coal business and never needed the Glamorganshire Canal again.

ENDNOTE

1 Extracts from Jeff Winter, *Taff Trail*.

Heading south-east, we arrive at the works of Brown Lenox and we can see the work's basin, locks 31 and 32 and the lovely old bridge that can still be photographed, but the locks have been partially filled in and fenced. This photograph is an RAF oblique, No.17824. The old Cardiff–Pentrebach road (the A4054) curves around the left and top of the shot, and Nightingale Bush, where the canal is still preserved from the bridge, which is in the car park of the Bunch of Grapes, for about half a mile south. Other than the Forest Farm at Coryton, this is the only preserved area of the Glamorganshire. Ynysowen House is at the bottom of the photograph. (Welsh Assembly Government Photographic Archive)

The bridge on the locks at Ynysangharad: the locks have been made safer with fencing and filling but they are still a hazard. The bridge forms part of the Nightingale Bush conservation walk, behind the Bunch of Grapes. (I. Jones)

This old photograph delivers a great moment in that inter-war period. Bill Bladen leads his horse Dick and the boat from the lock, with his son Bert at the tiller. It is 1935 and Bill has just finished discharging at the basin of Brown Lenox and will now make the journey south. (Author's Collection)

Boats leaving Ynysangharad Locks would follow a curve that bears left and brings you down to this position at Glyntaff, near the famous crematorium and St Mark's Church, heading for Gwern-y-Geryn. Pwllywhyad Bridge is a short distance further south. The railway station is at the left on higher ground. The Aberdare Iron Company's bridge is at the right. The Duke of Bridgewater's Arms is at the left of centre. St Mary's Church spire dominates the locality and a canal boat passes the wharf that once handled many tons of iron each day. All of this journey is now under the A470 and a 1970s-built housing estate at Rhydfelin. (Pontypridd Library)

This fine photograph of the canal in 1899 shows the canal bridge near the Llanbradach Arms in Gwern-y-Geryn. The workers, surrounded by stone, are on the canal company's wharf and the buildings are those of the company's workshops and warehouse. The stones are being assembled here in preparation for work on the bridge to begin. It was to be widened and strengthened for the electric tramways of Pontypridd to use its depot on the right (east) bank of the canal. (Pontypridd Museum)

This is 'Pont-y-Doctor'. Some called it 'Machine Bridge' because of the weighbridge on its end that measured the load of each tram that crossed over the Taff to the canal, either to Pwllwhyad bridge wharf, or later to the Doctor's Canal. (Pontypridd Library)

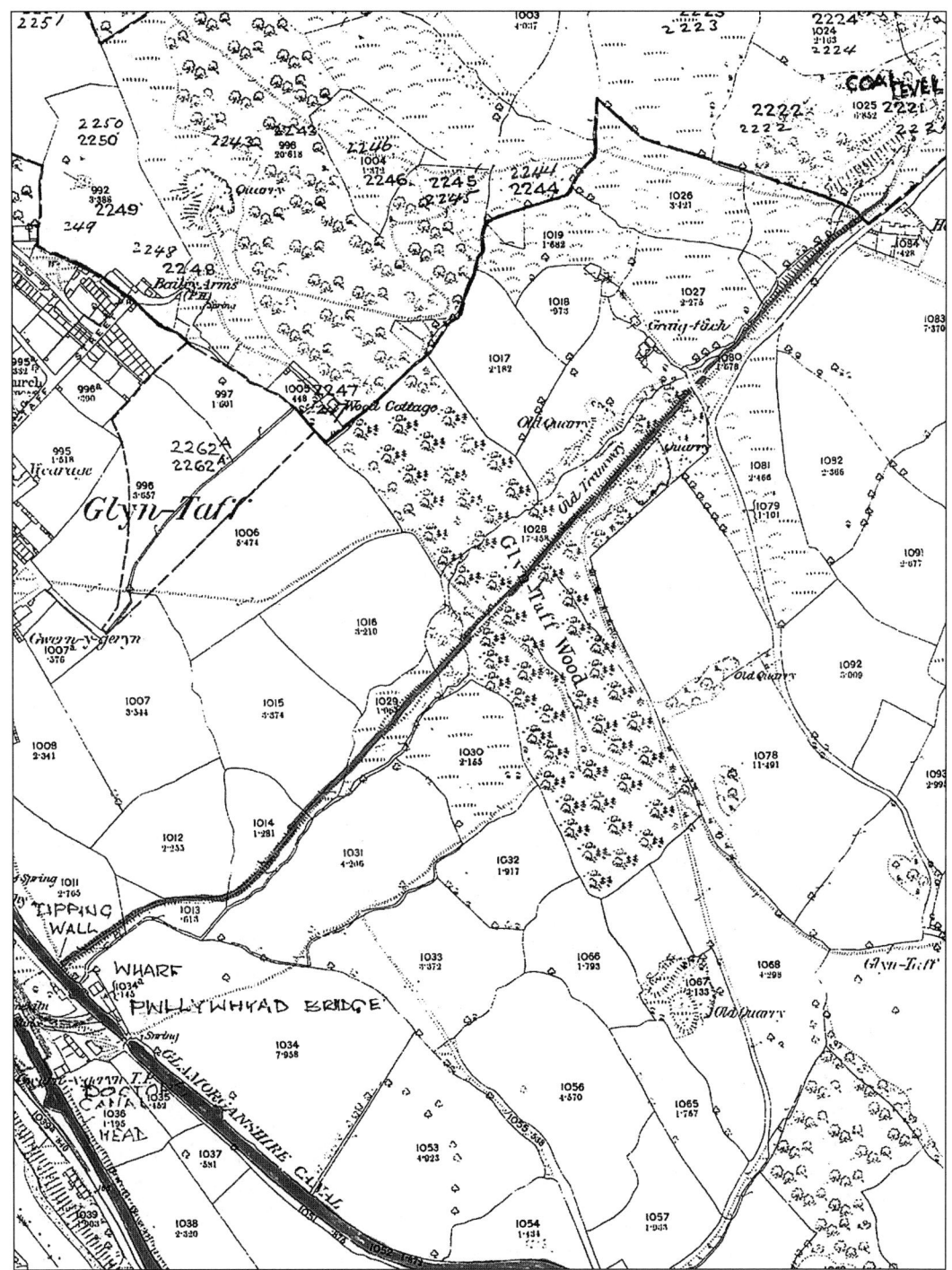

This 1874 Ordnance Survey map shows the long incline that was built for Bryn Tail Colliery. It is worth a visit to see what remains of the gullies and causeways that carried the tramroad up and down the mountain. Its purpose was to send a lot of coal to Cardiff via the wharf at Pwllywhyad Bridge, but almost everything that could go wrong, did. (Welsh Assembly Government Photographic Archive)

These are the remaining walls of the engine house that lowered and hauled trams up and down the hill at Bryn Tail Colliery. (I. Jones)

The tramroad had depressions and high ground to navigate and was built up into an embankment, which had been cut here to allow a path to cross through this gap, which was once a bridge (whose abutments can be seen). (I. Jones)

The canal and Pwllwhyad Bridge are long gone but this is where the bridge was, approximately where the two roads meet. A local man, Gareth Cox, has studied the canal locally and has been a great help in trying to position sites that are now under roads and housing. A street in Rhydfelin called Ilan Avenue has followed the course of the canal accurately until it runs out into school fields and then down to Upper Boat. In this photograph at the right in the distance, Ilan Avenue starts, almost under the A470. The small road in the left foreground takes you down the slope to the site of the Doctor's Canal head. This same slope is the one that carried railway lines from the Rhondda, over the Machine Bridge and up to the wharf. There was a canal basin here too, in the space between the blocks of flats shown. (I. Jones)

Moving south from Pwllwhyad we come to Lock Lewis or, properly, Dyffryn Lock and bridge No.33. This photograph dates from the summer of 1920, when it appears that a cruise party is about to begin. (Pontypridd Library)

Dyffryn Bridge, shown from its west side, carried the main road of Rhydfelin up the hill from the Cardiff to Pontypridd road. Dyffryn Crescent and Dyffryn Avenue are above. (Pontypridd Library)

About a mile south of Lock Lewis we come to Dynea Lock, No.34, with its lock house. The lock is at the left and the hill in the background has the name 'Monkey Hill'. This is where the Doctor's Canal made junction with the Glamorganshire Canal and this pool developed because of all the clay removed from here to maintain the integrity of the canal bed as it was leaking. Gravel was also extracted from here. Now it has all gone though one can still trace it. Housing has replaced part of the canal bed but it has mostly returned to nature. (Pontypridd Library)

A major breach in the wall of the canal occurred in 1908 near Dynea, causing much flooding and delay. The repair to the fabric of the canal was expensive, and this is where much of the clay was extracted. The lock house of Dynea is in the distance. (Cardiff Central Library)

This is an RAF oblique of 21 April 1943, a year after the closure of the canal. It shows the Taff Vale Railway from Rhydfelin at the foot of the photograph, under the wing tip of the PRU aircraft as it banks down to Upper Boat. At the foot of the photograph we can see that the canal still contains water and, at the wing tip, is the bridge of 'Ty Uchaf', then after that is the Dynea Lock where the Doctor's Canal joins. This canal, now filled in with ash from a nearby power station, can be traced back to the wing tip after passing the long row of houses. Further south is Maesyraul Bridge, then Foundry Bridge; we are at the spot where the canal turns eastward to Pentre Bridge. It looks as if Maesmawr Bridge is in sight where the road lifts in a squiggle above the three rows of housing. Running along parallel with the Doctor's Canal is the doomed Cardiff Railway just a couple of hundred yards away from its nemesis in the railway yard of Treforest. Further west (right) is the old Cardiff to Merthyr road and the Taff where the power station once stood. The bridge at Upper Boat is at the top of the photograph with the Treforest Trading Estate to the west of it. (Welsh Assembly Government Photograph Archive)

Julia Bridge in 1929 after suffering flood damage. The bridge provided the crossing of the River Taff to the Treforest Tinplate Works, enabling their finished plate to be loaded on to canal boats at Dyffryn Wharf. Raw materials came in the opposite direction. (Author's Collection)

This photograph, taken from the direction of Julia Bridge, shows part of the Treforest Tinplate Works, said to be the biggest in the UK for some time. (I. Jones)

Coal from George Insole's pits near the Taff, at the bottom right of the map, was brought by tramroad to join the main tramroad that ran from the foot of the incline at Ynys and ran across to the near bank of the Taff at Upper Boat where the amazing one-man ferry crossing took place. Maesmawr Colliery is at the left bottom and another level near Warren House above the main railway line. It happened that Maesbach coal also used this route to the canal. The Diwhyd Colliery Company's incline is shown at the bottom left at its low level. (Welsh Assembly Government Photographic Archive)

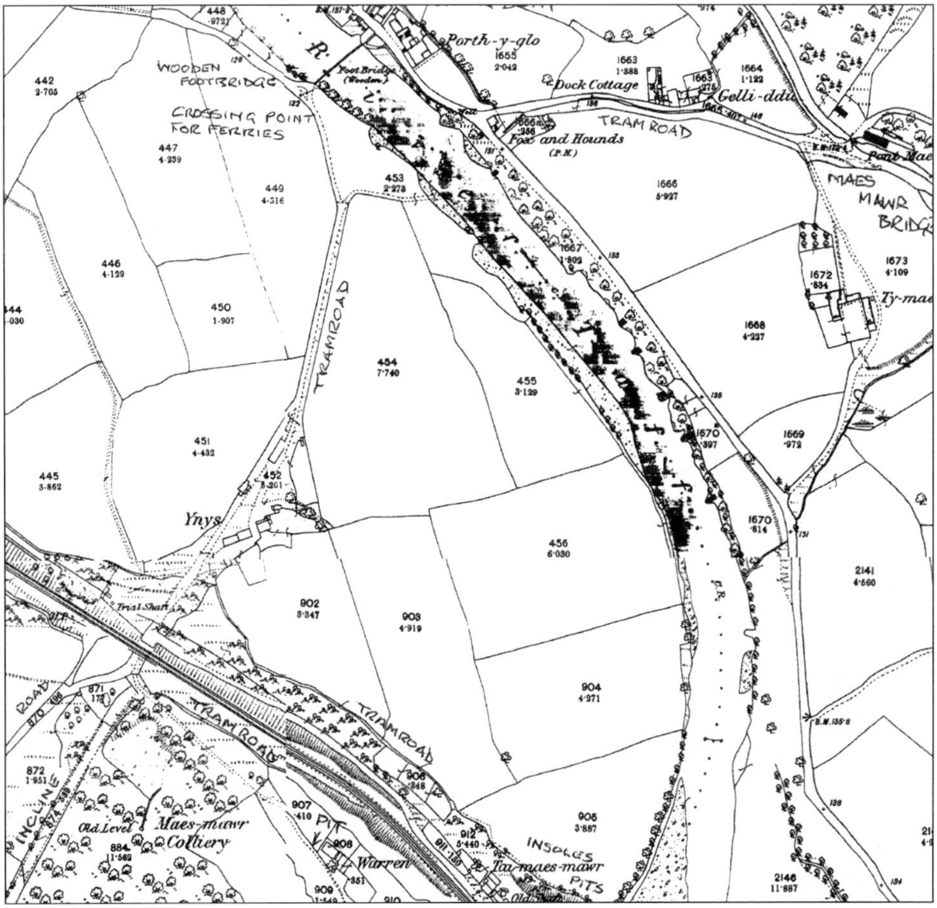

This part of the spoil tip at Maesmawr has a high proportion of coal. This photograph was taken on the tramroad from Warren Level. (I. Jones)

The tramroad to the railway bridge at the incline bottom from Warren Farm Level. (I. Jones)

This coal level near Warren Farm was only recently the focus of the National Coal Board when children were found playing about in this level and somehow brought about a fall in the entrance some 25 yards in from the portal. It is now faced in stone and secure. The letterbox slit was deliberately fitted to allow free movement of young bats that were born after the level was closed. Whilst the National Coal Board were performing this task, the rails of the tramroad were found under a foot of soil in the gully shown in the photograph. (I. Jones)

This was the original Taff Vale Railway Bridge. Well built, it survives today and is continually in use between Cardiff and Merthyr or Pontypridd. The arch on the left, in foliage, was that used by the tramroads from many pits. It was also serving the incline bottom, about 300 yards to the left and uphill. The road on the right comes from the Taff Bridge at Upper Boat and then climbs this nasty gradient up to Maesmawr Farm and the incline top, then on to Tonteg and Llantwit Fardre etc. (I. Jones)

This building was originally the incline top winding engine. Built for the Dihwyd Colliery Company, it is now a very impressive private residence. The winding was done at the ground floor level where the small windows are, and the height of the top floor was needed to see the bottom of the incline and signals from the house at incline bottom. (I. Jones)

After travelling half a mile eastward from the incline bottom and railway bridge we come to the Taff. There was no bridge in those days and a square, metal ferry boat would be poled across taking two trams of coal – a perilous job. To maintain a true course, a rope was stretched over the river above the ferry which was secured to the rope by guides fore and aft. On reaching the eastern bank the trams were pushed on to the tramway that led to Maesmawr Wharf. (I. Jones)

6

On Our Way to Cardiff

THE MAESMAWR INCLINE AND DIHEWYD COLLIERY

The famous Thomas Powell's Colliery at Llantwit Fardre in 1837 had no access to a railway or even to decent roads for his coal, but he did know of plans going forward for the Taff Vale Railway that was to run north–south past the foot of the hill at Maesmawr Farm. Powell planned to use the parliamentary 'four-mile clause' of the tramroad access to the Glamorganshire Canal which would give him compulsory purchase of the strips of land needed for a tramroad from his colliery at Llantwit Fardre to the Maesmawr tramroad that ran to Upper Boat, and at a cheap price. It was 1¾ miles to the incline house at the top of the hill from Dihewyd Colliery, then down the hill to the incline bottom, which was driven by an endless chain of one dram downhill lifting another uphill. There were many coal owners that guessed what Powell's intentions were, but most doubted he could flout the law and get away with it.

After arriving at the foot of the hill, his coal would be tipped into Taff Vale Railway wagons and, if the timing was right, none of his coal would go by canal as Parliament thought. He did get away with it. Most of the incline can be traced, as can the trek across to Llantwit Fardre. The incline top house is now converted into a beautiful three-storey dwelling and the incline bottom house is a respectable cottage.

Canal customers on the lower Taff west bank stop at this stage southward to Cardiff, as the pits and levels of Pentyrch and Gwaelod were not sending their coal by canal. The smallish quantities left, after the Pentyrch Ironworks had their first amount, was sold for domestic and local blacksmith consumption.

We have then reached the southward limit of the South Wales coalfield.

THE COLLIERIES AND CONSUMERS OF THE EASTERN SIDE

Moving downstream of Maesmawr Bridge on the canal, we pass a large turning place for the boats and then another bridge at Ty'n-y-wern Farm. This was to provide an outlet for the small collieries on the slopes of Mynydd Meio, to send coal to this wharf via their respective tramroads. The first of these, if we work southwards, was at the level of the Eglwsilan Road, where a shaft had been sunk on top of the mountain. A coal level and quarry were working downhill from the shaft and lying on the side of Eglwysilan Road about a hundred metres below the shaft. Further down the slope was yet another coal level covered now by reeds, characteristic of the acids at work in areas of land where coal is present. Lower down the mountain on the Fynnonbwla Road is Tair Level. The present farmer and other locals tell me that Tair Level means 'third level' and demonstrates the other levels as 'first' and 'second'.

Anyway, the evidence is there to see today, as it is also on the 1874 Ordnance Survey map. Who worked the levels is unknown; all that is known is that the Tair Level was working a seam called the 'Stinking Vein' of 24in thickness towards the end of the nineteenth century. Tair had a massive spoil-built embankment to run its tramroad on and it is still there today. The farmer showed me the mouth of the level near the road.

It appears that the tramroad ran to the Ty'n-y-wern tipping wall, but when the Pontypridd, Caerphilly and Newport line was built, it cut across the tramroad without a bridge being built for access. This suggests that the P, C & N, when completed in 1889, didn't have to build a tramroad as the mine was closed by then. The line of the tramroad south of the railway and the road 'Heol-y-Bwynsy' can be traced on the 1874 map and near the crossroads, though here it is on private land.

The next colliery south was 'Groes' or 'Pen-y-groes' to the north-west of Pen-y-groes Farm. The initial coal level that was developed here, just south of the Eglwysilan Road, was probably the one operated by the GCC's principal clerk, Mr Phillip Williams, who had a lease of twenty-one years from October 1805 but died in 1808. He built a tramroad to the Ty'n-y-wern Wharf, which incurred much outlay of money, and this money was in the form of a loan from Richard Crawshay, who stripped the place of its few assets until stopped by the law, in an attempt to get his money back.

Many attempts to sell the mine were made but nobody seemed interested, until along came Brockett Grover to buy the lease and the level with its fine, long tramroad. He got it back into production by January 1811. The 1874 Ordnance Survey map shows this level as disused. In 1860 another level was opened just south-west of Brockett Grover's and probably used the old tramroad. It was operated under the collective title of 'Mynydd Meio Collieries' and, in August 1863, the three partners, John Owen, E.C. Downing and Thomas Jones, were all shipping coal to the Cardiff Sea Lock from Ty'n-y-wern Wharf. The date of this company ceasing to work is not known but it was still active in 1874 as the Ordnance Survey map reveals.

Prospecting started in this area in 1890 once again and a level was developed at 'Fynnon Wen', which was very near to the last two mentioned mines, but much nearer the canal. Ironically, it was never to use it as the coal went to Newport Docks via the Pontypridd, Caerphilly & Newport line that had opened four years before in 1886. A tramway ran to the siding not very far from the Groes Wen Railway Bridge at 'Crossroads'. The stream called Nant Ffynnon Wen runs parallel with all the last three described levels. The name of this place, appropriately, was Groeswen Colliery, belonging to Beddoe & Co., which was another colliery working the Coed Cae Dyrys seam which had a section at this place comprising coal 30in, dirt 3in and coal 13in.

COED CAE DYRYS COLLIERY

Just half a mile south of Pen-y-groes Farm was another coal level up the hill in the forest of Coed Cae Dyrys. In August 1855 Evan Williams of Dyffryn Ffrwd struck coal in this wood north of Nantgarw. He laid a tramroad down to the canal, not to Ty'n-y-wern this time, as it had become such a long way to run a tramroad, but to a point a few hundred yards upstream of the Nantgarw Pottery Bridge. In 1858 he started to dig a level that extended under the Glamorganshire Canal. This caused much concern to the canal company, who sent a Mr John Lewis to examine what had been done, but the canal breached and lost its water. The workings of the mine were flooded. Williams had to pay the GCC £658 19s 11d damages. The Coed Cae Dyrys Colliery continued to bring coal up from the canal-side pit but made no attempts to go under the canal. A wooden bridge was made to pass the spoil across to the western bank as it was feared that spoil on the hill slope on the east bank would slide down into the canal.

The 1874 OS map shows the tramroad and the incline to the upper level as disused, but the level close to the canal was shown to be in use as the 'Coed Cae Dyrys Colliery'. It later became known as the 'Nantgarw Llantwit Colliery', which was abandoned in July 1893. It worked the Coed Cae Dyrys seam, which had a thickness of 18–22in. The site of the upper level is easily found, and the track bed of the tramroad, but it is believed that the electricity substation at Nantgarw occupies the site of the canal-side level. Nantgarw Colliery never used the canal.

The next customer of the canal on our way down to Cardiff was not a colliery, but a pottery.

NANTGARW POTTERY

Fine porcelain was manufactured at Nantgarw between 1813 and 1822, while a range of cheaper glazed earthenware was also produced through the Victorian era.

William Weston Young chose the site and started building kilns alongside the Glamorganshire Canal, which offered a cheap and easy means of transport. William Billingsley (1758–1828) arrived at Nantgarw early in 1813 from the Worcester porcelain factory. He made his home at Nantgarw House, which the former owner, farmer Edward Edmunds, had used as an inn. With a meagre capital of £250, Billingsley, with his son-in-law Samuel Walker, founded the Nantgarw Pottery or China Works and the enterprise suffered from financial instability throughout its life.

The GCC considered the wharf and bridge at Nantgarw as a strategic stop. They not only had the task of shipping the products of the works down to Cardiff but raw materials also came by boat; in addition all kinds of food shop goods came to supply the area. Many people at Nantgarw worked for the GCC, mostly as boatmen. There was a stopover for the night for the boatmen and their horses here and the canal bridge carried the road to Caerphilly. The canal company had a warehouse, blacksmith's shop and other houses that can still be seen today.

Coed Cae Dyrys Colliery shipped coal from the wharf at Nantgarw on an occasional basis.[1]

RETURNING TO THE COAL MINES ON THE CANAL AT THE SOUTHERN END OF THE CANAL

Graig-yr-Allt Colliery, Nantgarw

There were several attempts to expand this area of the coalfield just south of the pottery and the bridge at Nantgarw. The canal is on a ridge, high on the slopes of Graig-yr-allt Mountain, with the Cardiff to Merthyr road to the west and then the River Taff alongside that.

In 1795 a lease was acquired by William Lewis of Pentyrch Ironworks. The canal had been cut and the potential for shipping coal was obvious. He took a twenty-one-year coal lease at £10 per annum plus 6d per ton of coal. Others had cut coal earlier along these levels, such as William Key and his brother Thomas, later sending most of it to the Melingriffith Tinplate Works by horse and cart, joined by the ubiquitous Brockett Grover until the lease expired in January 1826. This was coal on Dyffryn Ffrwd land. The land that was leased to William Lewis was the property of John Goodrich of Energlyn. A few months later, Thomas Key took a thirty-one-year lease from Goodrich on neighbouring Graig-yr-allt land.

Graig Bridge on the canal was nearby and the GCC built a wharf on the downstream, eastern side of the bridge to facilitate the loading of coal. A Mr Morgan Thomas took up the challenge and opened another level on land belonging to the Williamses of Dyffryn Ffrwd and sent his coal the short distance by tramroad to Graig Bridge and thence to Cardiff until approximately 1850, when it seems to have been taken over by Richard Blakemore of the Melingriffith Tinplate Works, who had been receiving coal from Graig-yr-allt for many years for consumption at the works. It was in 1859 that the Melingriffith was given permission by the GCC to erect an incline bridge over the ridge carrying the canal to the Graig Bridge, all the way up from the level on the Cardiff Road. This powered incline was to be further extended to the east so that coal could be loaded into trucks in a siding of the new Rumney Railway of the Walnut Tree Junction branch.

The colliery closed down before 1878, but the exact date is unknown. The Graig Bridge, canal and everything appertaining to the canal are now under the A470. At the level of the A4054 (Cardiff Road), where the garages and offices are situated between the ridge carrying the canal and the Cardiff Railway embankment, is the office of the Ferris Bus Depot. The concreted wall of the ridge face by the side of the office is where the mouth of the level was. Witnesses tell me of the water coloured by iron and other chemical matter that used to leak from these banks.

This colliery worked the Nos 1 and 2 Rhondda seams at the end of the nineteenth century. On the 1898 map it is shown as abandoned.

Further South: Bryncoch Colliery

On the eastern side of the canal below the long-gone village of Maes Milwr is the combined brickworks and colliery where fireclay and ironstone was worked in the 1850s by John Edmunds, the son of Edward Edmunds of nearby Penrhos. Tests showed coal deposits within the quarries, so in 1852, the GCC gave John permission to build a quarry wall and passing place for boats on the east bank of the canal at Bryncoch. In 1861 the colliery suffered an inundation of water. In 1863, Thomas Williams of College Ironworks of Llandaff had taken over the brickworks and colliery at a dead rent of £75 per annum, a royalty of 4d per ton and a wayleave of 1d per ton. Four years later, in 1867, this Goodrich land was acquired by the Plymouth estate whose books show the half-yearly volumes of fireclay and ironstone raised – but not coal. The royalties on coal raised still went to the Goodrich estate.

The canal bridge at Bryncoch was busy as coal, bricks and clay all found their way to the canal. The Glamorgan Record Office has trade during the period 1864–67 in the hands of Dobson, Brown and Adams, but income was poor and they were soon in arrears. During this period, the material produced came to 11,582 tons of fireclay and 188 tons of ironstone. The coal side of the company was to pay royalties to the Goodrich estate.

Taffs Well Colliery

This was situated alongside the Nant Llywydd brook at a place of dense woodland, just east of the cemetery at Taffs Well. There is little to be seen of the coal level today, but the site is inside the fence near the brook on the east side of the road. The track of the old tramroad that carried the coal downhill to the canal crossed the railway at a crossing over what is now the Taff Trail track. This colliery was operated by and for the Brown Lenox anchor and chain maker of Newbridge for their Ynysangharad Works on the canal.

The tramroad ran down to the Glamorganshire Canal to a point just north of the Cae Glas Lock, where a wharf and a dock existed for boat repairs. The A470 cuts off the tramroad and the Cae Glas Wharf is now under concrete on the industrial park. This tramroad made junction with another tramroad from another level run by Thomas Thomas. The junction was made at the level crossing of the Rhymney Railway branch line.

Thomas Thomas Colliery

In 1878, Thomas Thomas, of the Bryncoch Colliery and Brickworks, also leased the Rookwood Colliery from the Plymouth estate. This lay above the Ty Rhiw cemetery and appears to be closely associated with the Bryncoch undertaking, which started twenty-eight years before the well-known Rookwood opened. The tramroad from this level could have been located on the north side of the cemetery wall, joining up with the Taffs Well Colliery's tramroad on the south side of the cemetery.

The Rookwood Colliery of 1906 was here as well as the 'Rookwood Slant' of 1914 and the New Rookwood of 1937, which closed in 1963. All used the Rhymney Valley Railway, later the Great Western Railway. None of these made any use of the canal.

The Garth Works Taffs Well

What a varied past can be seen in the history of these works. It was built as the Garth Anchor and Chain Works in 1864 on a site near the Walnut Tree Viaduct and alongside the Glamorganshire Canal on its west side. The works were blessed with a basin on the canal of a good size to accommodate several boats and a cast-iron towpath bridge to allow entry in or out of the basin. This was erected in 1865. On the opposite side of the canal, at this point, was the wharf and tramroad terminus of the Portobello Quarry.

The Brown Lenox Company had been forging the same products as those contemplated by the Garth Company and, with fifty years of quality and experience to compete with, whether that was its downfall is not known, but they went into liquidation in 1876. The works' manager, during the years of production, was Isaac John Rees and in 1875 the general manager was J.W. Morgan. The undertaking's registered office was 14 Marsden Street, Manchester.

Perhaps the following court case in 1875 tells us something about the company:

Civil Court, South Wales Circuit, Swansea

This was an action for wrongful dismissal.

Mr Gifford QC and Mr Hughes were for the plaintiff.
Mr Macintire QC and Mr Coxon were for the defendants.

This case occupied two entire days and involved, as the learned judge observed in his summing-up to the jury:

> A question of very great importance as well to the plaintiff, as to all others engaged in the same employment. The plaintiff has been connected with the manufacture of iron since he was ten years old and had invented, and patented, a form of anchor which, in point of utility, was a considerable success. In the years 1872 and 1873 he was concerned, together with some Gentlemen of eminence in connection with Naval matters, in promoting a company for the manufacture of anchors and chain cables. They were successful in forming a Company called the 'Garth Anchor Chain and Iron Company' and works were established in the neighbourhood of Cardiff. The plaintiff was appointed Manager of the works on a salary of £500 per annum which was to be increased in proportion to the increase in profits of the works until it should reach the maximum sum of £1,000 per annum. The appointment confirmed by a resolution of the Directors entered in the Minute Book and by an agreement between the plaintiff and the company under which he was to retain the situation for seven years. In July 1874 after some correspondence between the Directors and the plaintiff, in which the former made various complaints against his mode of conducting the management, he was summarily dismissed by them. He then brought this action for damages for the dismissal. The defence relied upon was that he had either appropriated funds of the Company for his own use or else by his negligence, suffered them to be appropriated by others, when, if he had performed his duty to his employers, he would have prevented it.

To prove this, the assistant manager, the cashier and a clerk employed in the works were called. The clerk stated that it was part of his duty to make out, every Thursday, a list of the number of days each man employed at the works had worked during the preceding week and the amount due to him for the week's work. This was done by reference to a book called the 'Time Book' in which an account was kept of the work done by the men during the week. He further stated that upon Thursday 28 May 1874 the plaintiff entered his office and, seeing that the amount payable to the men for the preceding week was much smaller than usual, told him to increase the amount by £50 by adding on days to the account of each man who had not worked the full week, saying that it was for a sum due for other matters and that he had the sanction of the directors for making the alterations. The witness accordingly made the alteration and the account showed the company to be liable for £75 for men's wages for the week ending that day, although the cashier received from the plaintiff only £26 and that amount only was paid to the men. The cashier confirmed this statement saying he was present when the instruction was given by the plaintiff to the clerk. He added that the plaintiff had, on the following morning, shown him his bank book and told him that the amount added was to recoup for himself a sum of £50 which he had paid as a gratuity to the manager of a neighbouring ironworks to induce him to push forward an order for cable iron which the plaintiff had given him, and £5 subscribed for an organ for a church in an adjoining parish.

The assistant manager, whose duty it was to sign the pay sheets, stated that he was surprised at the largeness of the amount charged for wages when called upon to sign that of 28 May; but upon being told that the plaintiff had said that it was satisfactory and that the greater part of it was for 'advances' to the workmen, he then paid the money to the cashier without any further investigation.

The questions left to the jury were: whether the plaintiff had caused the false account to be made out in order to convert the difference to his own use; and, if not, whether he was guilty of culpable negligence in allowing the account to pass without examining it.

The jury retired for an hour and then returned a verdict for the defendants (the company) upon the second question but dismissed the complicity of the plaintiff in the falsification of the accounts.

◆ ◆ ◆

When the Garth Anchor and Chain Company went into liquidation in 1876, its assets were sold at auction in 1881 to another anchor and chain manufacturer from Cradley Heath, but the premises were assigned to the Garth estate in the names of Frank Morgan and Joseph Gibbs in 1884, who commenced negotiations with a Parisian called Charles Audauy, who was to establish a Patent Fuel Works on the site. This particular industry had proved a winner elsewhere in South Wales, and the French had been in the forefront of using coal dust and small coal from here in making fuel blocks and briquettes. The important requisite was cheap transport; bringing the coal from the valleys and sending it down to Cardiff was to be initially by canal.

The Rumney Railway had gained the right to use the Taff Vale Railway's metals up as far as Walnut Tree junction at Taffs Well. The line of the RR left that of the TVR near the Garth Works and a siding was laid to the Patent Fuel premises so that the loaded wagons of fuel briquettes would move over 300 yards of RR track to join the TVR, whose metals would complete the journey. Considering that the RR was already a sore on the back of the TVR, trouble loomed! There were arguments and disputes between the two carriers until 1885, when agreement was reached which stated: 'All traffic from the works to Cardiff should be over TVR lines except 33% of the patent fuel manufactured at the said works that shall be sent by the Glamorganshire Canal Company.' Production of this ovoid-shaped fuel was to last from 1884 to 1900. In the 1891 census the manager of the works was Jenkin Owen, lodging at Church Street.

In 1901 the Garth Works underwent another change, when Thomas Gregory, listed in the 1901 census as a mechanical engineer living at the Foundry House near the Garth Works, took the factory on under the title of 'Thomas Gregory and Company, Iron founders'. Extensions were made to the works that doubled its length, and heavy machinery was also installed. The company flourished, making forgings, pressings and castings and, on the outbreak of the First World War, was awarded contracts for the War Office to provide nose caps for heavy artillery shells and shell casings, which were sent down to Cardiff by canal.

Somehow, the name 'Everard's' became associated with Gregory's, at least to the locals at this time, but no one could explain to me why! If a worker was asked, 'Where do you work?' the answer was often Everard's and not Gregory's.

The year 1926 was to see the end of both Thomas Gregory and his engineering works. His funeral on Monday 25 January attracted much attention from the press. It would appear he was a good employer and benefactor. In the upheaval that followed, speculation about work was the topic of the day. By the end of 1926 another company showed interest in the works but it was not for engineering. Zinc Oxides Ltd took on the works in 1926 and apparently stayed until 1938, though I'm afraid no one knows what they did there.

South Wales Forgemasters was established at Cardiff in 1938 and took over the Garth Works. Further improvements were made and new equipment installed. The company used road and rail but the canal was not to be used again at this basin; in fact it was filled in and built upon. After fifty years of good work, the company closed down in 1987, but after a short period the Welsh Development Agency gave enough money to enable part of the works' business to survive. At the time of writing, Forgemasters are still on the site, but many of their buildings are being used by other firms such as Horseley Bridge Tanks, who make assembly kits for large industrial, water or other liquid tanks. Other buildings are being used by car dismantlers.

◆ ◆ ◆

This section comes from a small book written by a working man, Donald Gordon Bunn, who lived his whole life in Taffs Well, and was given to me by his son, Gordon Bunn.

I have extracted the following passages from his book, which also discusses religious, social and sporting aspects of life in the area:

Memories of Taffs Well

When you hear this tale you may wonder why I have written it. Being born in the village at 6 Yew Street, my grandfather living in Treble Locks and having three boats on the canal, my wife's grandfather being at Walnut Tree, thereby having close connections with the top and bottom of the village, I feel free to write them. So I shall start with the canal.

My grandfather, John Fraser, lived at Treble Locks and worked three boats between Cardiff and Pontypridd with the help of his three sons, namely Tom, Johnny and Stanley. One boat was contracted out to Hopkin Morgan's Bakery, Pontypridd, one for general cargo and one to the Crown Fuel Works, Maindy, Cardiff. Opposite the Treble Locks is the reservoir, or 'big pond', which supplemented the canal during the dry summer months. We boys practically lived there during the summer holidays. It was reckoned to be one mile around. To deter us, old Roger Williams of Rhiw Ddar Farm used to keep the bull in the field that we had to cross to get to the pond.

Now the loading bays on the Glamorganshire Canal at Taffs Well started at Treble Locks. Treble Locks had vehicle access from the main road across the Cardiff Railway, and then there was Thomas's Lock. There was access there for vehicles with general cargo, then we arrive at Cae Glas Lock and there we find the tramroad from the level at Cwm Brynnau all the way past the cemetery and alongside the brook to the canal.

We next arrive at the Garth Works, otherwise known to the local people as 'Everards'. It was there that they had a basin or pond in the works ground and that between the 1914 and 1918 war, they made shell cases that were loaded into the canal boats in this basin. The entrance to that basin was under the hump-backed bridge opposite the place known as 'Canal Row'.

Then there was also a quarry known as the 'Portobello' which brought stone down to the canal between those houses already mentioned, Canal Row and Forest Row.

Now we arrive at the Walnut Tree and, therefore, we get to the Walnut Tree Bridge where there was erected a wooden Gantry for loading coal onto the barges from the Rumney Railway.

[It is the first I had heard of this gantry!]

There was also, a little lower down, the Castle Coch Quarry with access to the canal, which was underneath the Cardiff Railway and on to the wharf on the canal where stone was loaded on to the boats. Now the portion of the canal above Treble Locks is the longest of the canal without a lock. Known to the boatmen as the 'three mile pond', in that stretch we have the Nantgarw Pottery, we have Nantgarw Colliery, we have Dai Gash's Knackers yard (or slaughterhouse), then there was also the Bryncoch Brickyard, Bryncoch Colliery and there were several other places.

Having travelled down the canal, we shall travel back up the road through the wonderful village of Taffs Well. Let us start at the Walnut Tree Viaduct. A wonderful piece of workmanship, the foreman mason lived in No.4 Anchor Street, Taffs Well. His name was Harry Morse, who could boast he knew the number of bricks in each pillar. This piece of work was not without tragedy, namely Mr Sloper who lost both eyes and arms in blasting operations, now ending up sitting under the Rumney Bridge in Cardiff with a metal cup.

There is much more of this delightful book of Taffs Well but it would take up space that would be inconsistent with this book of canals.[2]

◆ ◆ ◆

Passing now to the west bank of the Taff, we visit an ironworks that had no business with the Glamorganshire Canal, other than its ongoing battle to hold on to the water bled off the Taff that had always driven the blast house's wheels and the forge's rolls. When the canal opened, it drew off water from everywhere possible to maintain a height of water in the canal to enable a 20-ton load for each boat to pass. Another reason to visit this historic ironworks is its connections and trade with the Melingriffith Tinplate Works, who did have business with the canal, from the beginning to the end.

ENDNOTE

1 Extracts from Dan Powell, *Victorian Pontypridd*.
2 Many thanks to Mr Gordon Bunn for the use of these extracts.

This is the canal wharf at Pont-y-Glyn or Maes-yr-aul Bridge. The workmen are filling the canal boat with clay to repair the walls and floor of the canal at the breach at Dynea. This is probably the wharf used by Dynea tramroad to carry coal to Cardiff. (Pontypridd Museum)

This is Foundry Bridge Wharf, which transferred cargoes from Cardiff towards the breach at Dynea. These could go no further as the canal was dry between Dynea Locks and the stop dam above Maes-yr-aul. Boxes of fruit and groceries were carried by horse and cart, as can be seen. (Pontypridd Museum)

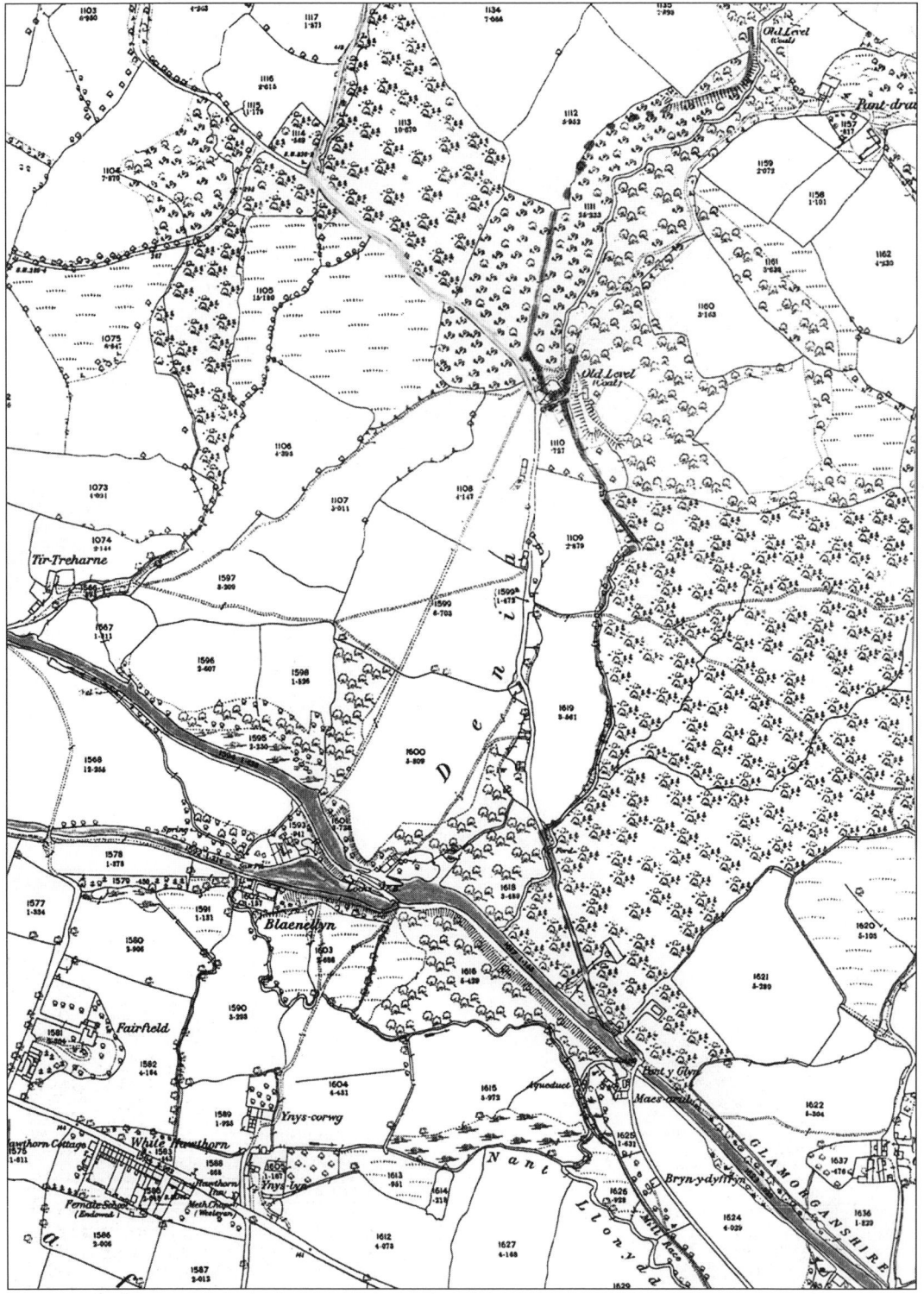

Pant Drain Farm, with the upper level at the top right. The map of 1875 shows the very busy Nant Corrwg travelling down through the impressive ravine to the junction of tracks, then to the farm main gate where there are two more levels. The upper and lower levels had tramroads that joined here and a single track is thought to have made the journey to Maes-yr-aul. The junction of the Glamorganshire and Doctor's canals at Dynea Lock can be seen on this map at lower middle. (Welsh Assembly Government Photograph Archive)

This fine photograph of Maesmawr Bridge was taken in 1954. Fifteen years later it disappeared under the A470. Originally known as Weaver's Bridge, it became so busy with the coal from across the Taff (that is Maesmawr and Maesbach) that it was renamed Maesmawr Bridge. It had a boat dock on its eastern side where loading of coal for Cardiff could take place without obstructing other passing boats. (Mr D. Chaplin)

This is Tyn-yr-Wern Farm Bridge. Tramroads from the several small collieries on the slopes of Mynydd Maio would lead to this bridge, where a coal-loading wharf was built. No sign of this remains now as the A470 runs down close to the left-hand side of the photograph and the university playing fields run along the right-hand side. The bridge also had the name of the nearby road – 'Heol-y-Bwynsy'. (Pontypridd Library)

This RAF overhead, dated 31 May 1963, shows many interesting things on the eastern Taff Vale. At the bottom right (the dotted line) is the canal, covered with trees. The '+' mark is the site of Tyn-yr-Wern Farm, on the west side of the canal. The broad line that runs parallel to the canal is the railway once known as the Pontypridd, Caerphilly and Newport line (later Alexandra Dock and Railway). The bridge over this line is at a crossroads where the several tiny, narrow roads ran to many coal levels and pits on Mynydd Maio. In the 1800s most of these coal mines had tramroads running down to the canal bridge at Tyn-yr-Wern Farm. The clear track of the tramroad from Groes is in the middle. At the top right corner of the photograph is the conical spoil tip at Coed Cae Dyrys Colliery. Other locations shown are: (1) the spoil tip at Groes at 'Ysgybor Wen'; (2) Tair Levels tramroad; (3) Groes Level; (4) the curving road to the coal shafts on top of the mountain; (5) a coal level and a stone quarry; (6) another coal level, only shown by reeds growing in a spray pattern; (7) Groes-Wen Colliery. Left is the way to Upper Boat and right is Nantgarw. (Welsh Assembly Government Photographic Archive)

Looking down from the coal shaft on the top of Mynedd Maio, one can see the coal level and quarry on this side of the Eglwysilan Road and on the far side of the road opposite is the area of reeds over the coal level that existed there up to a hundred years ago. (I. Jones)

The old canal bridge at Cae-ty-du is left between the A470 and a business park, deep in almost impenetrable jungle. It's been partly demolished, for reasons unknown, with huge pieces still lying in what was the canal bed. (I. Jones)

This is probably the tramroad access built by the Pontypridd, Caerphilly and Newport Railway. This rail bed is now the Taff Trail cycle route. The tramroad has descended down from the Coed Cae Dyrys coal level into the underpass, before terminating on the east canal bank at Nantgarw. (I. Jones)

This shows the track downhill that once carried the tramroad from the site of the coal spoil tip (right) of Coed Cae Dyrys Colliery, halfway up the slopes of Mynedd Maio. (I. Jones)

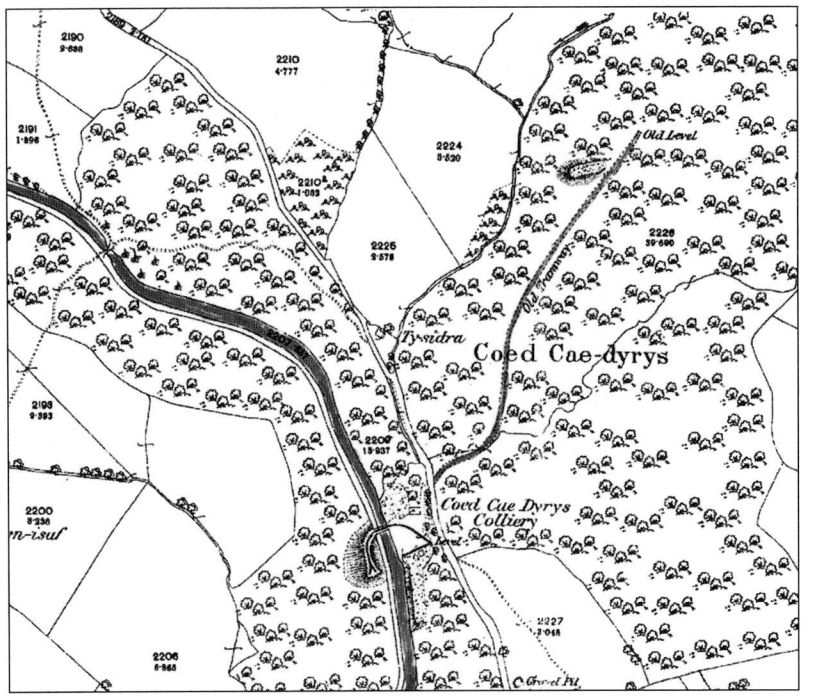

Coed Cae Dyrys Level near Nantgarw in 1874, at the top right of the map, and its long incline tramroad down to the Glamorganshire Canal. The loop over the canal was a wooden bridge built for the purpose of sending colliery waste and spoil over to the lower slopes of the hill on the west side of the canal. (Pontypridd Library)

Nantgarw Pottery, on the north side of Nantgarw Bridge, photographed from a canal boat in 1890, with another canal boat lying alongside the wharf. Many boatmen lived in Nantgarw and sometimes as many as twenty or thirty might lie overnight after journeys from both directions. (Taffs Well Library)

The old bridge at Nantgarw, viewed from the south. There were many men from this village working for the GCC. (Taffs Well Library)

The GCC's former warehouse, alongside the wharf at Nantgarw. This is a lovely spot high above the village, with fine views. The trickle of water that represents the line of the canal is fenced off for safety reasons, but the old bridge and wharf do give one a sense of this once-busy place. (I. Jones)

Moving south on a raised ledge above the Cardiff Road, on the outskirts of Nantgarw, we reach the coal level at Graig-y-Allt (shown on this 1874 Ordnance Survey map). The canal runs north to south down the middle of the map. On the left are the main Cardiff Road and the old Cross Keys Inn. Graig Bridge is below centre of the map with the Graig-y-Allt incline elevator crossing over the canal just south of the bridge. The Rhymney and Barry lines are at the bottom right and the actual Graig-y-Allt Colliery is at the bottom left. It was close to the road and was closed due to flooding many times. The River Taff was only a short distance to the west and that would flood too. There is now nothing to be seen of Graig-y-Allt or the canal as the A470 runs over the site. (Welsh Assembly Government Photographic Archive)

This is the site of the large coal level that ran along the roadside of the Cardiff Road. This is the car park at Ferris Coaches Tour Office. (I. Jones)

Graig Bridge facing north. The Graig-y-Allt Colliery used this bridge and wharf on the downstream, east side of the canal, until the construction of the incline that took coal over the canal to a siding on the Rhymney Railway line where the coal was tipped into trucks. This bridge was built simply to serve the people on the east side at Tair-y-Graig who otherwise would have lost connection to Nantgarw. Mr Morgan Thomas opened another level on land belonging to the Williams of Duffryn Ffrwd and he sent his coal to the wharf and then Cardiff via a short tramroad until 1850. The bridge and canal are under the A470 now. (S.C. Fox Photograph, Cardiff Central Library)

RAF overhead 58/676, 12 May 1951, No.3435. The village of Nantgarw is at the left of the picture. The junction of the Cardiff to Merthyr Road and the road to Caerphilly, as it was then, is evident.
  (1) is the site of the Graig-yr-Allt Colliery and incline (Nantgarw); (2) is Graig Canal Bridge; (3) is old Rhymney Walnut Tree junction line; (4) is the Barry Railway; (5) is the Treble Locks, Nos 35, 36 and 37; (6) is the GCC's reservoir for the canal's water level; (7) is the River Taff; (8) is the Cardiff Railway; (9) is Cardiff Road; (10) is Bryncoch Colliery; (11) is the Bryncoch Canal Bridge; and (12) is the Bryncoch Brickworks site. The canal was dry and gradually being filled, mostly with colliery spoil. It is nearly ten years after closure of this canal. (Welsh Assembly Government Photographic Archive)

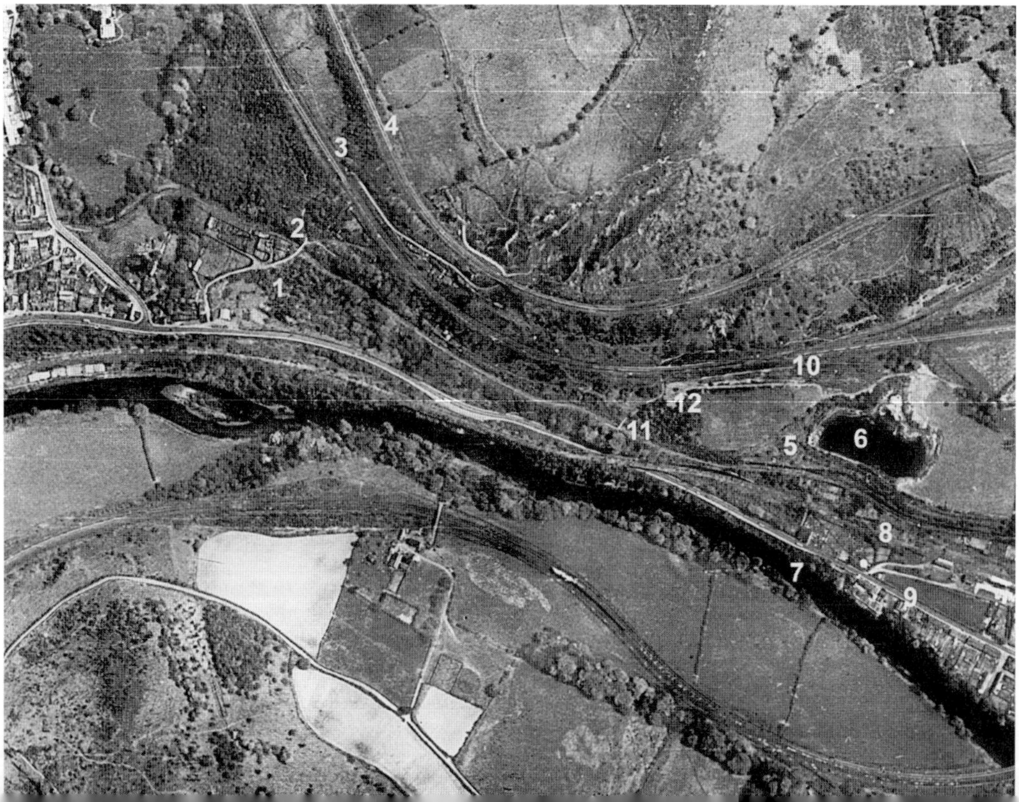

Above: The famous Treble Locks, Nos 35, 36 and 37. The smaller building behind the two lock houses is the boatman's rest and the buildings survive, although the rest house is engulfed in dense vegetation. The A470 runs over this site and is now alongside the house wall on the towpath side. (Cardiff Central Library)

Above right: Here are Treble Locks cottages today, in a lovely spot, although next to a major road. The boatmen's rest house is still in the back garden of the nearest house and, on the land some way from the front of these cottages in the undergrowth, one can find the sluices of the old canal that cleared excess water from the canal right through to the River Taff. The excellent masonry and brickwork of the GCC are still here. (I. Jones)

Tracing south, this is the bottom lock and the canal proceeds to Taffs Well. The company's reservoir is on the left. (Ian L. Wright Collection)

This lovely Victorian postcard was meant to advertise the Graig Mountains, but gives us a good view of Taffs Well Lock 38. The building in the foreground is a boatman's shelter and stable at the other end. The lock house has recently undergone an expensive facelift by the Rhondda-Cynon-Taff Authorities. The present incumbent, Mrs Nancy Dimond, was the wife of one of the lock keepers. (Taffs Well Library)

This is the refurbished lock house and the last of the few inhabitants, Mrs Nancy Dimond. There are still signs of lock masonry and iron fittings in the grass. The spot inspires thoughts of the past, as do Mrs Dimond's tales of the days when the boatmen shared the rest house with the horse. (I. Jones)

A short distance downstream from Taffs Well Lock, we come to Cae-Glas Lock, No.39. A wharf or tipping wall was situated a few yards upstream on the north side of the canal. This was to be the end of a tramroad that ran down from the slopes of Mynedd Maio. There were several small levels on the hill and the main tramroad brought coal down from Taffs Well Colliery which, at one time, was leased by the Brown Lenox Chain and Anchor Works at Ynysangharad. It was the case then that coal mined in Taffs Well was being boated on the canal upstream to their works. Another, to the north of this, was believed to be that of Thomas Thomas Colliery, that lay on Plymouth estate land near to what became Rockwood Colliery. This ran down a track now used as a bridleway down the northern side of the cemetery and joined with the Brown Lenox tramroad before going under the Rhymney Railway Walnut Tree junction line, then headed for Cae-Glas Wharf. Most of this can be traced today as footpaths, but Cae-Glas Lock and House are now under the concrete of the Taffs Well Industrial Estate. (Pontypridd Library)

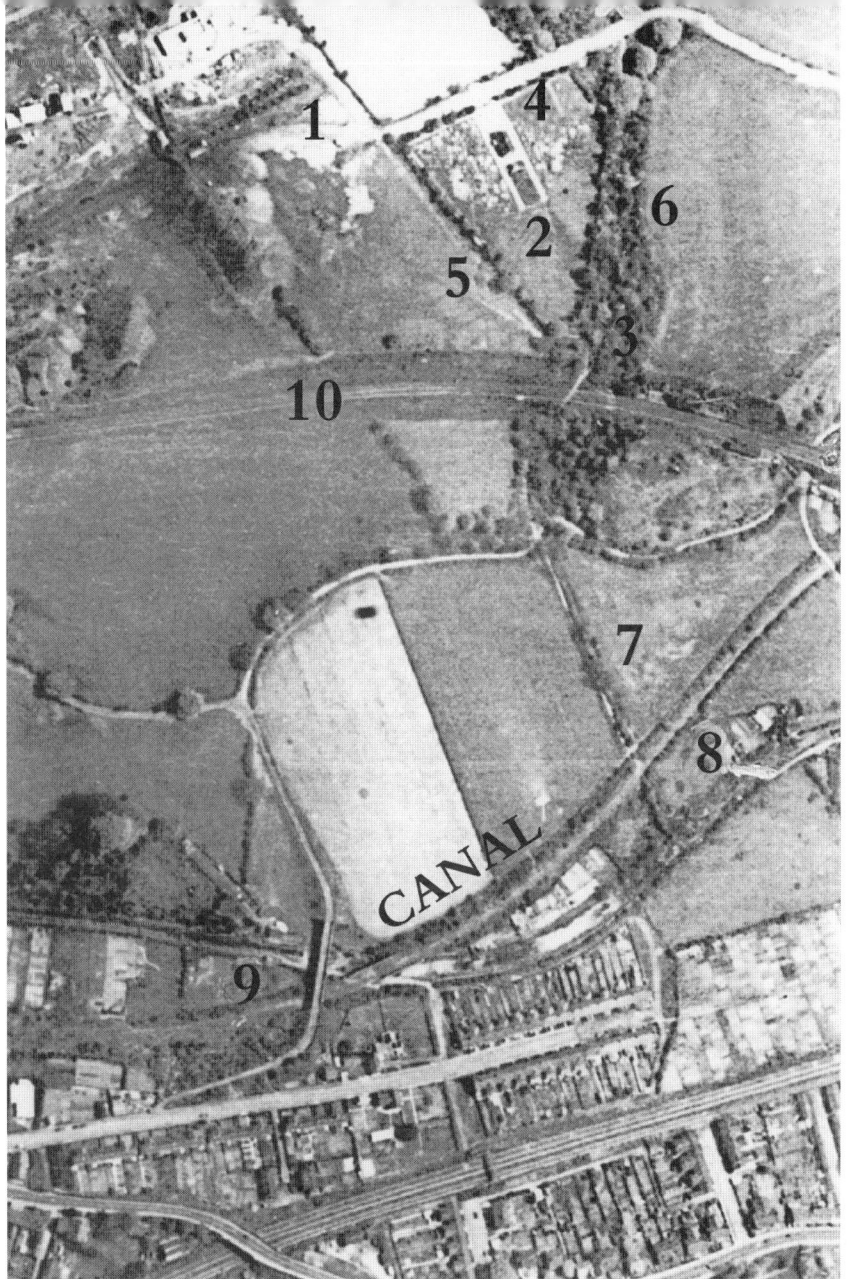

Another RAF photograph showing the area around Taffs Well. The now dry canal curves to the east from these locks to pass under the line of the Rhymney Railway Bridge at the middle right (out of picture). The Brown Lenox Level is at the top of the photograph marked with a '+'. The village of Taffs Well is at the bottom of the photograph. (Welsh Assembly Government Photographic Archive)

(1) The believed road level of Thomas Thomas Colliery
(2) The Ty Rhiw cemetery
(3) The junction of the two tramroads can still be seen under the trees
(4) The Cemetery Road
(5) The colliery's tramroad
(6) The junction with the Brown Lenox tramroad
(7) The road to Cae-Glas Wharf
(8) Cae-Glas Lock
(9) Taffs Well Lock
(10) Rhymney Railway track

This is the middle part of the tramroad from Taffs Well Colliery. (I. Jones)

Portobello Lock, No.40, now under an industrial estate at Taffs Well. To the right of this shot is the northern end of a terrace of houses called Forest Row and, to the rear of that, can be seen the stone quarry of Portobello. (Ian L. Wright)

This 1900 photograph shows us moving in a southbound direction. We pass under the Walnut Tree Viaduct of the Barry Dock and Railway Company, with the River Taff on the left beyond the hedges and trees. The Taff Vale Railway line is clear at the left, then the canal, then at the right the Cardiff to Merthyr Road. The Cardiff Railway was also at far right, out of shot. At the right side of the canal can be seen the tipping wall or wharf of the quarry nearby. (Taffs Well Library)

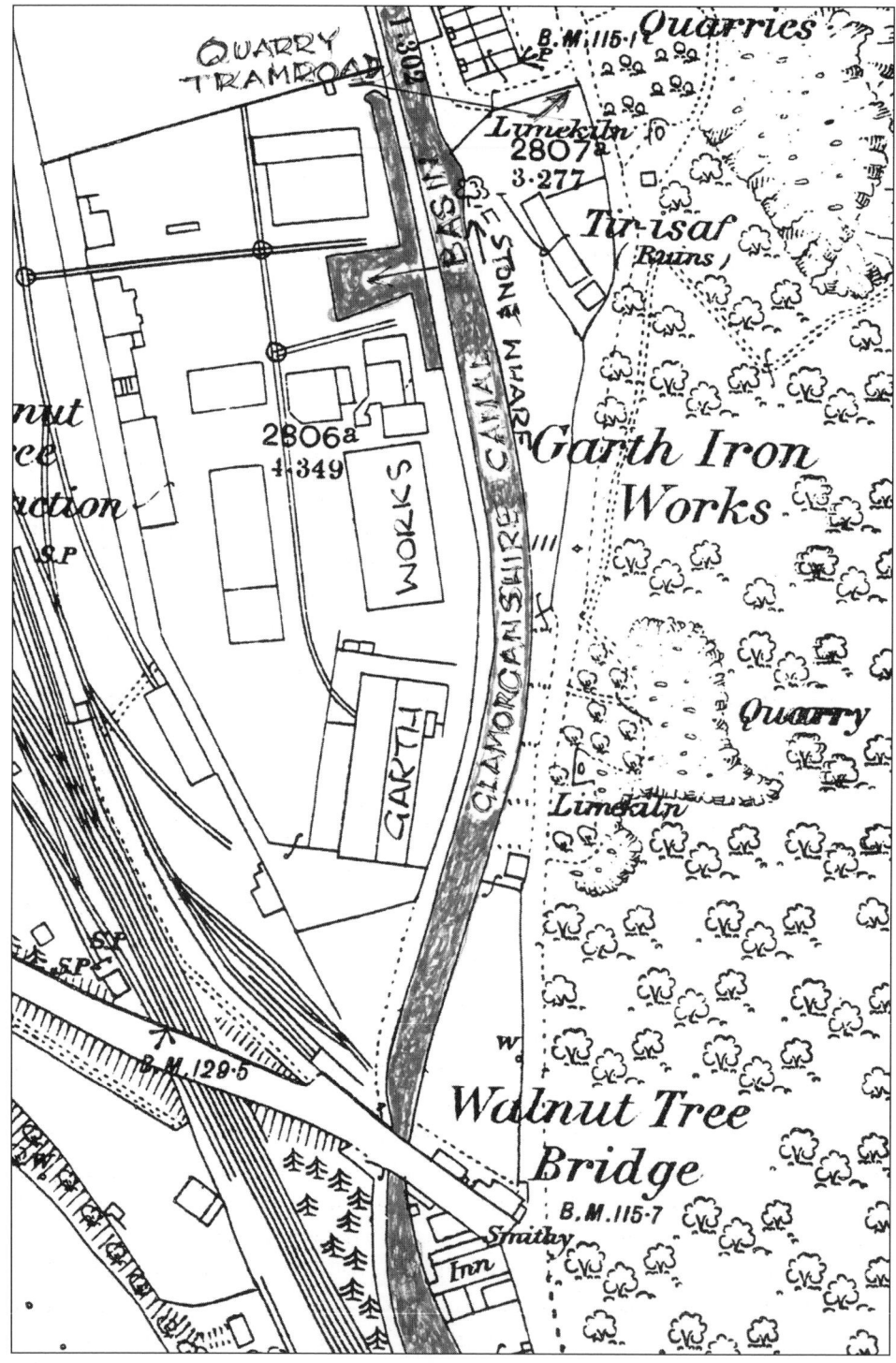

On the opposite bank to the Portobello Quarry was the Garth Works. This 1875 Ordnance Survey extract shows the canal running alongside the works, with the towpath that used to be carried by a cast-iron bridge to span the entrance of the work's own basin. There was a canal wharf opposite the basin on the east bank which was used to ship stone brought down from the quarry at Portobello. Limestone was the product with two limekilns shown on this map. (Welsh Assembly Government Photographic Archive)

7

Pentyrch Ironworks

The parish of Pentyrch lies on the west side of the River Taff a few miles north of Cardiff, where the coastal plain gives way to the hills of upland Glamorgan, and was the southern limit of the coal measures. The lower Garth Hill here contained a mass of haematite iron ore that was surrounded by limestone, necessary as a flux. The hillsides were well wooded and on the eastern side of the Taff the woodlands of Fforrest Isha, Fforrest Fawr and Fforrest Ganol were more than adequate to supply the charcoal required for the fuel of those early ironworks. There was an abundance of water in the area to supply the blast furnaces, and sandstone, to provide building material, was found on Garth Hill.

This combination of easily worked resources was noticed in the sixteenth century by some of the Sussex ironmasters, under pressure to conserve the now-scarce timber resources of the Weald, as they had used up so much, not only for charcoal, but for boatbuilding and other uses. They came to South Wales, and in the years 1564–68 Edmund Roberts was active in developing the ores of the lesser Garth to supply the Robertsbridge forge in Sussex, but this was a long trip and was uneconomical. By 1574 a furnace and gun foundry were worked by members of the Edmund Matthews family who were the landowners. They made guns and sold them illegally to the potential enemies of the state, probably exported from Cardiff Old Quay. By 1625 the furnace was closed down by the government and, after an appeal to Charles I that failed, the place was demolished, both furnace and forge.

In 1800 there were two furnaces called the 'lighter' and 'lower' furnaces run by the owner, Mr Lewis, and his manager, Mr Vaughan. New furnaces for melting and puddling were introduced. They produced wrought-iron bars from a mixture of charcoal and pig iron. (Lewis was a major partner in the Dowlais Ironworks.)

This mixed iron was named 'tin bars' and were sent to the Melingriffith Tinplate Works for rolling into tinplate. In 1805, Lewis sold the works to the Harfords who owned the Melingriffith, thus linking the two neighbouring works. In 1810, the Harfords sold the combined works to their cousin Richard Blakemore, who had been studying the trade for some years.

The Blakemores, and later Richard's nephew Thomas William Booker, retained the companies quite successfully and prosperously until the economic condition in early

1882 forced the closure of the Pentyrch Works after almost three centuries of iron making.

These ironworks and forges had been blessed with their sources when the transition from charcoal came about. Nearby Lan Colliery provided them with coal for coking and for their steam engines. Half a mile north of the Lan Colliery, a new seam was opened called the 'Cwm Dous' (Sweet Valley) Level. By 1973 the total output of the Booker collieries was said to be 100,000 tons per annum. This total would be increased to 160,000 tons when the new openings were worked. These new workings were the 'New Lan', which opened in 1875 just 20 yards north of 'Old Lan', and the 'Sidrig' Level, which opened in 1878, a mile north of Old Lan. The clearance of the site has been complete, beginning as early as the 1890s when the slag heaps were removed to provide ballast for the construction of new docks at Cardiff.

THE PENTYRCH TO MELINGRIFFITH RAILWAY

Richard Blakemore built a tramroad between 1812 and 1815 which followed the River Taff and works feeders and crossed the Taff at Gilynys Farm Bridge. In 1871 the Melingriffith Company converted its tramroad to a standard-gauge railway and built an iron bridge to replace the wooden one at Gilynys, and a spur to the lock on the canal at Tongwynlais. This railway then went through the ironworks at Pentyrch and up to a connection with the Taff Vale Railway at Ynysgau.

THE MELINGRIFFITH TINPLATE WORKS

A Brief History

The works here are probably the oldest industrial site in Wales when we go back to its very beginning. I thank Edgar Chappell for his research on this typically excellent work – 'Historic Melingriffith'.

It is first mentioned as a manorial mill of the Lordship of Senghennydd, and at a later date, a mill for the Manor of 'Album Monasterium'. The name Melin Griffith (Griffith's Mill) was probably derived from Gruffydd, son of Ivor Bach, who was Lord of Senghennydd during the latter part of the twelfth century. This mill was driven, throughout its existence, by water directed along an artificial course from the Taff at Radyr Weir, and this method of obtaining power was in operation for a part of the works until the closure, when most of the plant was using steam engines.

The first references to ironworking here was in 1749, with a lease from Mr Rees Powell. Richard Jordan and Francis Homfray took this lease of twenty-one years to a corn grist mill called 'Velin Griffith' and a forge in the Parish of Whitchurch. The Jordans disposed of their interest to a Bristol firm called Reynolds, Gently & Co. in 1765, and among the partners of this firm was an ironmaster of the name Harford who was a well-known Bristol Quaker. By 1786 the ownership was in the hands of Harford, Partridge & Co. – a firm who were to be active at other South Wales ironworks.

In the early years of the nineteenth century a newcomer arrived at Melingriffith, Richard Blakemore, who was the nephew of John Partridge, and was earmarked for great things. He was learning the trade and managing another of the company's concerns at Monmouth Forge, but taking a real paternal interest in the Melingriffith Works. He joined Harford and Partridge at their legal battles with the GCC's attempts to reduce the factory's water supply. By July 1808, the partnership of Harford and Partridge Company had been dissolved. Richard Blakemore, with Joseph Reynolds and two of the old partners Thomas Pritchard and Richard Jones-Tomlinson, took over the Melingriffith Company which, since 1805, also included the Pentyrch Ironworks. The new company was Reynolds, Blakemore & Co., and Blakemore took on the management control. Over the years he acquired the interests of his fellow partners and became the complete master of his trade. He brought the Melingriffith up to scratch in every way, building the railway from Pentyrch to the Melingriffith, and opened new mills in the works. He also opened more small collieries in the area to supply his furnaces. The company mostly received coal via the canal from nearby collieries such as Graig-yr-Allt at Nantgarw. Blakemore also acquired the experience of the endless battles with the GCC, and the Harfords and Partridges, becoming a formidable opponent.

Blakemore was a bachelor and in 1820 he adopted a nephew, Thomas William Booker, who, in the meantime, succeeded to the business trading as T.W. Booker & Co., which prospered. T.W. Booker's three sons were given an interest in the company. The company then experienced financial difficulties and sought outside capital. Outside directors joined the board of what was now T.W. Booker & Co. Ltd, with the second son T.W. Booker as managing director.

At this time the Melingriffith Works comprised: twelve mills; two forges; cold rolls; and a shearing machine, all powered by ten waterwheels. The three forges at Pentyrch were powered by five waterwheels and achieved an output per annum of 10,000 tons of sheet iron and 100,000 boxes of tinplate.

Pentyrch became a liability due to the advent of steel as a superior metal for tinplate and the cost of converting the works to the Bessemer process was impossible due to the exorbitant cost. The works were closed as the sale of iron was falling everywhere.

The customers of Melingriffith were failing to pay their debts and the company's bank, West of England and District Bank, failed; as a result, a winding-up order was made against it. In July 1881 the Booker undertaking was put up for sale as a going concern but there were no bidders.

The company was ultimately leased to the Cardiff Iron and Tinplate Co. Ltd, with Mr James Spence as managing director, and the Pentyrch section was closed. The Melingriffith operated only intermittently and in 1887 it suspended operations with the Spence Company, going into liquidation.

In June 1888 the Melingriffith and Pentyrch properties were sold in lots by public auction. The Melingriffith Works and Railway, along with thirty-nine freehold cottages, were bought by Richard Thomas of Lydbrook for the low sum of £12,000, plus £10,500 for machinery and plant. A new company, the Melingriffith Co. Ltd, was formed with a capital of £40,000, with Richard Thomas as managing director. The new company began favourably, profiting from the cheap acquisition and the advantage of water power over their competitors. It was the biggest tinplate works in Wales and under the skilful management of Richard Thomas the works were reorganised and improved and he remained in charge until 1916, the control passing to his sons. Mr Spencer Thomas became managing director when, in 1934, the works were sold to Richard Thomas & Co. Ltd (no longer the Melingriffith Co. Ltd). Production at these works was suspended during the Second World War and the works became a virtual warehouse for the Ministry of Defence.

Post-war, after reopening tinplate production, it was absorbed into the Steel Company of Wales and the quality of the product kept them very busy until the late 1950s, when the continuous strip mills at Velindre and Trostre opened, which closed down all the hand-rolling mills in Wales. I am led to believe that Melin was the last of these to close. The site of the factory lay undisturbed from about 1958 until the early 1970s, when the heavy contract plant moved in and demolished it, and it is now a fine housing estate alongside the Taff.

THE MELINGRIFFITH'S BATTLE FOR WATER

When the GCC built the canal it was confident that the water supply to the canal would be adequately catered for in their plans, but it did not work out that way and, considering that all ironworks along the canal were using water power to run their works, it was also difficult to quantify. The GCC had failed to allow for losses of water at each lock movement; the water in the canal was also leaking through the bed and walls, and there was also evaporation to consider. The GCC was dictated to by Richard Crawshay of Cyfarthfa Ironworks and his attitude was always one of selfishness. Many of the ironworks had long legal fights with the canal committee over water but none with such regularity as the Melingriffith Works, whether under the Harford or Blakemore regimes. There was, to give one example, a meeting of the committee members responsible for the lower canal on the matter of: 'The erection of an engine at the tail of the Melingriffith Works mill-race, for the purpose of conveying water into the canal, and the proprietors of the Melingriffith Works agreeing thereto.'

The canal committee comprised: Richard Crawshay, Richard Griffiths (of the Doctor's Canal), Jon Bassett, Joseph Vaughan and Watkin George. The plan was to pump water from the tail-race of the Melingriffith feeder and channel it into the canal at the Melingriffith Lock. This same feeder had also been used as a canal by the works until the tramroad was built to bring iron bar from Pentyrch Ironworks to Melingriffith via the Taff and a feeder.

Canal boats would leave a wharf at Pentyrch and head downstream for a mile to the lock at the mouth of the feeder which was just above Radyr Weir. The lock house was still there up until the 1980s. Half a mile, at least, brought the boat to the Melingriffith Works.

This story is true but it amazes me that it was done. A powerful river and only poles to navigate the bends must have been even more dangerous than the Maesmawr tram ferry at Upper Boat – they made 'em tough in those days!

Of course, this was not the main use of the feeder. It was in continuous work as a supply to the waterwheels and turbines of the mills. In the summer months, or during any extended dry spell, the works would shut down and the men would be sent home. This is one of the reasons for anger with the GCC.

The Melingriffith Company agreed to the proposed water pump: after all, if the water being displaced from the tail-race was going to be pumped into the canal, it didn't bother the Melingriffith Company as it had always been sent to the River Taff after leaving the works.

William Lewis of Pentyrch Ironworks was to get his chance to put his argument before the Melingriffith started their rows with the GCC

In 1802 he blamed the lack of water in the Taff in dry season on the GCC building their second feeder at Quaker's Yard. These two feeders to the canal swallowed all the water of the Taff, even in moderate rainfall. He asked the court:

> Did the canal company think they owned the Taff and its water? The canal proprietors were illegally granting leases to themselves so that works could be supplied with water from the canal. At Nantgarw the Parliamentary Weir, where water should be continually flowing over, there was none. So, no water entered the Taff there. The reason was because, at Treble Locks, the GCC had built another weir at the head of these locks with a watercourse around the locks and into the pond below. This weir was four inches lower in the crown than the Parliamentary lock and, not content with that, they had added a sluice gate in this weir that is nine inches high and three feet wide, so not a drop is left to go back to the river and our works.

Lewis won his case and was awarded £100 damages, although his claim was originally for £1,500.

Harford and Partridge then prepared their own action, bringing the young Richard Blakemore, who was learning the trade. After more complaints from the Melingriffith, and to avoid more visits to court, the canal company proposed that:

> A reservoir should be made below Treble Locks capable of containing 500 locks of water and that such reservoir should be discharged during the course of the working week to maintain the lock's water levels for the navigation from the Treble Locks to the canal pond below the Melingriffith lock and that a waste-weir should be made below Treble Locks to discharge all the water taken into the canal (except Sundays) into the River Taff above Pentyrch Weir. A side paddle should be put in at the lock below Treble Locks for the purpose of discharging the water thereof into the aqueduct to be made for conveying the water from the said lock into the River Taff.

What this meant was that the GCC recognised the water problem for the twin companies of Pentyrch and Melingriffith and a reservoir would store excess water flowing around the canal on Sundays (when Pentyrch mills were standing) so that it could be used by the canal during the week. This would then allow Pentyrch use of the waters of the river during the working week.

With regard to the Melingriffith pump, it recommended that the channel from the waterwheel of the pump to the river, where waste water from the mill-race ends up, should be deepened. This would enable the canal company to lower the waterwheel by 2ft and so remove the necessity for the sluices which currently dammed back the tail-race from the tinplate works and interfered with the effective operation of the works' machinery. Mr Blakemore, for the Melingriffith, accepted these recommendations and, prior to the presentation of the report, had offered to pay one half of the expenses of the alteration at the wheel and watercourse to the expense to him of £200. Everyone was content, so Blakemore withdrew his latest legal proceeding.

There were many more shortages of water at the Melingriffith, but the canal was also suffering from water shortages here. Boats were grounding below Melingriffith Lock, at Cathays pound, and also at the tunnel below Crockherbtown – which was the last lock before the sea lock at that time. In the summer of 1826, its lock keeper, Thomas Collins, recorded boats that needed re-floating there almost every day – in one day there were fifteen boats that had to be eased off the canal bed by Collins raising the top paddle of the lock to let water through the tunnel. The company installed a stop gate at South Gate Bridge (Custom House Bridge) to allow the water between it and the tunnel to be maintained at an adequate level. Even ships in the sea lock pound were going aground.

In June 1828 Blakemore was withdrawing water from the engine (waterwheel) at Melingriffith, and this time it was the GCC's turn to threaten litigation against Blakemore. His answer was that 'the canal company had no right to enlarge the cut so many years after such powers of its Act of Parliament had lapsed' (the canal company was digging deep into the bed of the canal to make it deeper because of the groundings, but this increase in depth meant that more water was needed to increase the volume).

Judgement was in Blakemore's favour. In 1834 in the courts, Blakemore won his cases again and again and, at last, the GCC gave in, agreeing to pay compensation for water taken as long as he, Blakemore, agreed to allow the expansion of the canal and the increase in traffic to proceed. As a result of this, the GCC was to pay Blakemore £1,500 per annum from 25 March 1834 to 25 March 1839. In return, the canal company was entitled to take up to 10 tons of water per minute past the Pentyrch and Melingriffith works in times of short water without paying any additional consideration to the extent of 15 tons. Any additional use was to be paid to Blakemore at a rate of £100 per annum.

After this legal warfare and other cases afterwards, this old warhorse retired back to Monmouthshire. The management of Melingriffith passed on to his adopted nephew Thomas William Booker-Blakemore. Richard Blakemore died in 1855 and, although Thomas Booker had attended many of the trials and court hearings with his uncle, he did not continue the legal struggles with the GCC as the company he now ran went over to power by steam engine, leaving only one turbine to power the mills. In total, his uncle had spent over £20,000 in litigation.

The old pump that caused so much trouble can still be seen near the spot where the lock was. The Oxford House Industrial Archaeology Society of Risca and members of the Inland Waterways Association spent their spare weekends refurbishing the pump in 1975. The job was completed most successfully.

My Experience of the Melingriffith Works

I write this only to show what it was like during the period 1950–53 trying to keep the mills going here, while knowing that the enterprise did not have a future because of the modern, continuous strip mills being built by the Steel Company of Wales at Llanelli. The Melingriffith still had a good name abroad, as well as in the UK, and I was given the impression that order books for tinplate were still good.

I am eighty-two and I assume that, as I was a young man when I worked there, that there are not many men left to write about those years at this old works. That is the reason I include this in this story of the Glamorganshire and Aberdare canals, as a look back at old industry.

I was a chap that served his apprenticeship with the Great Western Railway docks division, later to become Docks and Inland Waterways and, although there was plenty of work about, I needed plenty of overtime as I was married with two children. It was through a third party already working there that I first heard of the conditions and earnings enjoyed by skilled men on maintenance there, and arranged an interview with the chief engineer. I joined the company at twenty-three years of age, at a time when the old machinery needed a lot of work to keep things running. Even the fitting and machine shops were still equipped with belt-driven machinery from one big electric motor.

The work that the handful of fitters did was not only dangerous but dirty and the present-day safety laws would have seen the place closed down. The chief engineer was a great chap, as was the manager of the works, and the staff were a good bunch to work with. But you did not know what you would encounter each day that you clocked in. Sometimes, if the night shift working anywhere in the works – on the rolling mills, the gas plant, the boilers or engines – had a problem, one of the men would get on his bicycle and knock on your front door; it was call-out time. As I was a Whitchurch resident, I was the nearest.

The canal still wound its way around the walls of the works and the long basin inside the works was still in water, but the canal and basin, of course, were inactive – the canal was dead ten years before. Funnily enough, one of my tasks there was to build a steel bridge over the canal basin as the engineers' office, stores, blacksmith's and fitting shop were all on the north side of the basin and the works were on the south side, so that everybody had to walk along this long basin to get to the works. The bridge saved so much time that it was surprising that it had not been done when the basin was made redundant.

The one job that I remember so well took the whole spring and summer of 1951, I think. My chum Lionel, with a young summer student on one side, spent days under the works, with me and another student, in the damp, dark feeder course stripping down a huge water turbine with a view to overhauling it. It was made of brass and bronze and dated back to about 1820, with dozens of louvres that would be opened and closed by a wheel up in the control room. All these louvres had a brass shaft for a hinge. All these shafts, about two dozen of them, had been ordered from an old blueprint from the original manufacturer. An equal number of the louvres were renewed in cast brass because of wear in the hinge bores. All the control shafting had to be built up by welding and then turned in the biggest of the lathes as the shafts were about 15ft in length. The journals on these shafts had to be turned after the welding had built up the worn parts, but as it was not steel but wrought iron, the weld was not as good to get a finish. All this expense and trouble was worth it. This turbine had only been turned over once since the war ended and apparently was little used during the 1930s as it was clapped out. The steam engines in the mills kept the mills attached to the out-of-work turbine. The men who had worked here for many years had never seen this massive flywheel move at all.

I believe this project to overhaul the turbine was to show fuel efficiency to the parent company, the Steel Company of Wales, and when it was ready to run, my chum and I were given a soft Sunday job of observing the start-up. All the summer under the works was spent behind a dam preventing the feeder from drowning us, so the carpenters had cleared it. The top engine man, who was old enough to remember driving it many years before, was in the control room with the works manager, the chief engineer and a few other suits. We were on the mill floor making sure that the driveshaft to each mill was greased and there was no clutter around the rolls. There was the sound of a lot of water crashing down past us. We looked at the 40ft-diameter flywheel (a third of this flywheel's diameter was in a trench close to the wall of the mills). There was a judder for a few seconds and the slow rotation as the driver felt his way with this now powerful machine. No furnaces were lit, there was to be no clanging of steel bars, but the old flywheel was turning faster and faster. By midday, when we finished work, the manager said that the next shift to start was to light up the furnaces on the turbine mills and produce tinplate. We were told that the trials showed 95 per cent on the gauges and a job well done.

There were so many interesting jobs, such as building a 35ft-square Braithwaite water tank up over the roof of the boiler house (four Lancashire boilers), way up on four old cast-iron pillars that originally held up the roof of some old demolished part of the works. Then we burnt down, with oxyacetylene torches, the original, cylindrical, 150-year-old water tank that was now surplus to requirements, burning it into small pieces then throwing them to the floor from approximately 30ft. No scaffolding, just a ladder. My mate and I burnt that big old tank down until my mate had to go as the last piece for me to burn off was the piece I was standing on.

Another job entailed working on the electric extractor fan way up in its frame in the roof over the pickling tank of the annealing shop. The fumes from the acid in the tank played havoc with any steelwork and where it was drawn by fan into the tresses of the roof, the effects were damaging to RSJ angle iron, corrugated roof sheets and especially to the fan housing. It was liable to fall into the acid below. Access to the roof, to make good the damage, was by ladder to the roof joists, where we had to walk the plank to the point under the fan, then up a shorter ladder to reach the trunking around the extractor fan.

This maintenance programme each day to keep the tired old place working was almost written in the Bible. To many men at Melingriffith, it was a form of dedication. Whenever I pointed out that a task was too dangerous without the expense of scaffolding and other safeguards, I was ignored. There were half a dozen gantry cranes that regularly needed road wheel bushes. The steel keys holding the gear wheels to their shafts would come loose but the repairs were done where the crane driver left it, never over a designated stage at the end of the track that I proposed: a simple stage of angle iron and timber so that, if you fell, you didn't need to be killed.

As I've said, the Health and Safety people would have had to build an office here for long-stay occupation given the issues they would need to address, but it was, other than that, a very good place for a young fitter and turner to spend some years. It not only had the effect of sending you home each day pleased with what you'd achieved, but you also knew you had done a job that would not normally be yours anywhere else. We stepped over the line in many ways that would not be tolerated elsewhere without a strike. If they valued you as a craftsman they looked after you. I remember being asked to work late several nights in the week to turn parts for the new works at Velindre. I completed this work and my boss said that they were so pleased at the quick delivery that they bought a new lathe to install in our workshop. Early on, the orders were that nobody else but me should use it. One can only guess what that would have done anywhere else. The time flew whilst I was there, and I enjoyed my three years in this old museum of a place. It is now demolished, with a fine housing estate occupying the place alongside the Taff.

THE BRICKWORKS AT LLANDAFF NORTH

Solomon Andrews, the Cardiff businessman, had a vast range of enterprises, including buses, taxis, ironmongers, building and property. According to the list of ventures and business interests listed in an agreement of partnership between Solomon and his son Emile on 1 January 1884, one of the items was that of farming the lands of Ty-Mawr Farm in Whitchurch. This meant that they had bought the lands prior to 1884, which included the brickworks on that land. As he was a builder of large office blocks and many of the existing buildings in Cardiff city centre, he employed hundreds of people and never used any contractors or suppliers, but stocked his own materials or made them. It is not surprising that he refurbished this old works to make bricks for his own use and, no doubt, for sale on the west bank of the canal north of Gelli Bridge and south of Primrose Hill Bridge. The works were surrounded by clay pits that were supplying the works but we do not know if any materials were brought by canal boat or even whether the finished product was distributed by canal. Indeed, Andrews may have only manufactured brick for his own use. In 1889 a new stack was built, 110ft tall, and, at this time, the kiln engine, boiler and machinery were valued at £1,500. The brickworks were sited at the northern end of a strip of land which ran alongside the River Taff as far south as Bridge Road and as far north as the Taff Vale Railway line at Primrose Hill. The works were on the west bank of the Glamorganshire Canal, just a short distance north of Gelli Bridge. The surrounding land belonged to Ty-Mawr; in a surveyor's report of 1886 it appears that Solomon Andrews bought 40 acres at a probable cost of £7,000. Planning permission had been obtained for two streets and an application had been submitted for several building plots.

The two streets were built, Mary Street and Solomon Street (later renamed Andrews Road). Houses in Bridge Road were also built at the same time. The land between Mary Street and the River Taff was later sold to the Corporation at a reasonable figure for use as a public recreation area. More land was donated by Councillor Hailey and Emile Andrews. It was named Hailey Park as a compliment to the councillor. A bus depot was constructed in Andrews Road, which later became a match factory and later still a laundrette. So we have all this building going on and Solomon with his many involvements had his brickworks (the bus depot he built was for his buses), but how much was the Glamorganshire Canal involved?

All we know is the canal continued along the east side of his property until the ending of canal usage in the 1950s.

LLANDAFF YARD

The same length of the Glamorganshire Canal along what later became Hailey Park was, 100 years earlier, a busy canal community. Known as Llandaff Yard from the beginning of the movement of canal boats on the way from Merthyr to Cardiff, the place in the area of the public house called the Cow and Snuffers grew into a well-known wharf. This was a public wharf, as opposed to those belonging to industrial concerns that appeared all along the length of the canal. It was very much like the stop at Trallwn in Newbridge, now Pontypridd. Its first and foremost claim to fame is that of farm produce and its long-term delivery to the ironworks and coalfields of communities lining the canal. Only a small amount of hill farming existed up in the valley so food, vegetables and fruit from the Vale of Glamorgan and local market gardens supplied the swelling crowds of people moving into the Merthyr area.

At the time there were few good roads, mostly turnpikes: the road that runs from Llandaff City – splitting, at the toll gate on Llantrisant Road, with Bridge Road (now the A4054) – crossed the Taff then carried on over the bridge at the canal lock No.45 near the Cow and Snuffers, then on north to Pontypridd and Merthyr. This also gave access to the fertile farms of the Vale of Glamorgan from Llandaff City.

The enormous and growing population of Merthyr (it was far bigger than Cardiff and stayed that way for nearly 100 years) needed shop and market goods, hardware, timber, clothes, flour, grain and animal feed, as there were hundreds of horses employed in the ironworks and, later, the collieries. The first people to spot this need were the Key brothers, Thomas and John, whose other brother William was already a carrier by road (horse and cart or mule train) before the canal was built, so they were quick to use this new form of transport to service their markets. Thomas concentrated on mineral movements of coal and limestone and in 1795, the year following the opening of the canal, he leased land above the lock at Llandaff Bridge from Lord Llandaff and laid out a coal yard (this would be the area of the James and Jenkins Garage).

Hardware and furniture, at the start, all came from Bristol as there was little manufacturing in Cardiff then. The boats would have regular crossings from Bristol, which had a wharf dedicated to shipping to Newport and Cardiff. The ordered merchandise would arrive at the sea lock of the canal and transfer to canal boats, be it piano or wardrobe, for up-canal transportation. John Key, in the same year, had a partnership agreement with William Taitt of Dowlais Works to supply merchandise to Merthyr from his farm and grist mill at St Fagans. They had a warehouse at Merthyr managed by a Mr John Brown.

The GCC provided a public wharf to encourage this unexpected source of custom in 1798. In the first full year of the canal opening, the GCC show in their books that Thomas Key was delivering coal to Llandaff Yard from his collieries at Maesmawr and Abercanaid totalling 85.5 tons and taking 532 bales of straw back to Merthyr from Llandaff (the straw, of course, for the horses). John Key was sending flour, bran, butter and malt to the Merthyr warehouse at lock No.1.

John Steel, Rees James and Lewis Thomas were also sending material between Cardiff and Merthyr and occasionally used Llandaff Yard. In a month, Steel took over a ton of sacked flour from Llandaff to lock No.1. Rees James took flour to the Dowlais Wharf in Merthyr, whilst Lewis Thomas took malt, flour, potatoes, apples and pears to the same destination.

An indication that not all manufactured tools were imported from Bristol was an entry in the GCC book of 8 September when two gates, with four posts and a plough and harrow, were loaded at Llandaff Yard consigned to Dowlais.

In September 1798 the Pendarren Works delivered 10 tons of bricks and John Key delivered 17 tons of building stone from Pentyrch to Llandaff Yard. This pennant sandstone developed into a regular cargo for Cardiff during the expansion of the town.

Benjamin Hall of the GCC leased a wharf here in 1803 and many others were thinking of doing the same. It was going as well as Trallwn at Pontypridd so the canal company, on 7 December, ordered that 'a small dwelling be erected there' (the public wharf) at a cost of no more than £35. In 1807 the canal company enlarged the wharf and in 1814 they extended the warehouse. In these early years wharfage income at Llandaff Yard was significant enough to be itemised separately in the published GCC accounts, alongside the sea lock dues and sundry rents and tolls at the aqueduct (Abercynon). It appears that Llandaff Yard was the company's only public wharf.

In 1809 the road from Llandaff turnpike to the Merthyr turnpike gate in Whitchurch became an important throughroute and was turnpiked in 1826. This, with the work going on at the wharves, created a need that Evan Llewellin satisfied by opening his house for the sale of beer. In the old Harrison's map of 1830, it is shown as the Red Cow, but soon afterward it became the Cow and Snuffers, where many drunks met their end in the locks. Other signed beer houses close to the canal were the Exchange, the Pineapple, the Gardner's Arms, the Three Cups and the Roller's Arms. The last two were at College Locks and served the College Ironworks at the next lock downstream.

It is not surprising that many small industries were born from the sides of the canal. David Evans' foundry in 1861 was successful on a site near Key's former coal yard, now in the hands of Mr Grover and Mr Coffin, to continue to supply House and Smith's coal. For those with long memories who know of Irwin's car part shop, that is where, approximately, the Eagle Foundry was.

The River Taff Bridge was built in 1808. Before that, a ferry crossed the river. Ynys House is at the right centre. This bridge also crossed the Taff Vale Railway and the canal. The Cardiff to Merthyr turnpike road is at the far side. Castell Coch is at the top right and the quarry is at the left. (Glamorgan Record Office)

The Ynys Bridge has been changed now with the A470 dominating the Garth Gap. This is what it looks like today. The slimmed-down railway that used to comprise six sets of track is now a mere two (in the trees at the left). The Taff is even further to the left and there are no signs of quarrying, no canal and only one of the square pier structures of Walnut Tree Viaduct remain, at the left. (I. Jones)

The photographer is taking a photograph of Castell Coch (just out of shot upwards) and a canal boat has rounded the bend and is approaching Ivy Bridge. The nosy youngsters get as close as possible. (Whitchurch Library)

We are looking at the downstream flank of Ivy Bridge in Tongwynlais. The GCC's lengthsman lived at Ivy Dene at the extreme left. (Cardiff Central Library)

What a fine setting for a photograph or drawing! This is lock 41 at Tongwynlais, in 1910, with Castell Coch above. The lock-keeper's cottage is the white one facing us at an angle. The space on the right was the lay-by for the canal's weighbridge, before it was moved down to Kingsway at Cardiff. (Cardiff Central Library)

This RAF overhead photograph is very good in identifying salient points of interest of the canal's course and connections to other historical gems. (1) is Ton Lock; (2) is Gylynis Farm with the road running diagonally across the land to join the main line (running vertically at the left), and steam from an empty train visible heading north on the main line to the coalfields; (3) is the iron bridge with its rail connection east to Ton Lock and south to Melingriffith; (4) is Llwynmellt Farm, now demolished and under the M4/A470 roundabout at Coryton. This farm name was the official name given to the lock, known more to the boatmen as Middle Lock; (5) is that lock; (6) is the elegant three-arch stone bridge built by the Cardiff Railway Company at Coryton (this is still in place near the ASDA supermarket); (7) is the lock house at Llwynmellt; (8) is the waste water from the lock and (9) is Forest Lock and House. (Welsh Assembly Government Photograph Archive)

Llwynmellt Lock or Middle Lock in the 1930s. At the left is the Longwood that extends all along the conservation area of the canal. (Welsh Industrial and Maritime Museum)

A very rare look at the lock-keeper's cottage at Llwynmellt Lock. It was demolished many years ago. (Whitchurch Library)

Walking south along the pretty, preserved part of the canal between Middle and Forest Locks. (I. Jones)

Forest Lock, No.43, and bridge, seen in the 1930s. Because of its overflow weir, the house was virtually an island and, although the house has been demolished, the waterways have been retained. (Glamorgan Record Office)

Continuing south past Forest Lock we reach even more lovely canal scenes with rare birds and fish. (I. Jones)

This unique cast-iron bridge is spanning the Parliamentary Weir. It allowed excess water, which flowed over a sill, to flow into the feeder (above the bridge), thus helping to maintain a reasonable head of water for the mills at the Melingriffith Tin Plate Works. This towpath bridge, manufactured in 1851, can still be seen. The canal is filled in from here. (I. Jones)

Radyr Weir was built to supply a consistent intake of water from the Taff for the Melingriffith feeder. The house in the photograph on the east bank is the lock house that was the dwelling of the man responsible for some unusual duties. It seems that, before the tramroad was built to bring iron bar down to the Melingriffith Works, it was brought down by boats powered only by poles and muscle, entailing not only loading iron but navigating downstream in all kinds of conditions – two men risking their necks to not only reach the locks at the weir but going through the lock and poling the half mile to the bar shed at the works. How these intrepid men poled their way against the current on the return trip is unimaginable. The tramroad replaced this voyage and, in turn, the railway of the Melingriffith Company succeeded the trams. Note the Taff Vale Railway train at the left and on the hill at the right is Castell Coch. A simulation of the sluices at the entry to the feeder can still be seen and water still fills a part of the cut, but the house and locks were demolished some years ago. (Whitchurch Library)

Opposite: The Melingriffith Tinplate Works on 23 July 1942, RAF oblique. It was not in production at this wartime period as it was acquired by the Royal Navy as naval stores for the duration. The River Taff is at the right, sweeping round the bend. The tall stack on the river bank was part of the boiler house where five Lancashire boilers delivered the steam pressure for the many steam engines that drove the rolling mills. The long shed flanking the river bank is the rolling mills for the black iron to be rough-rolled and cut. The stack to the left of the boiler stack is the annealing shed stack, and the pickling plant is between the two. The two long sheds alongside the canal were for cold-rolling and finishing (right), and inspection, packing and delivery (left). The long shed in the centre was the bar shed and cropping. Then, further up, were the scrap presses and benches where women separated the black plate into eight sheets to be pickled, annealed and finish-rolled. Coming through the works' entrance at the middle left, we can see the canal going under the works bridge, then a concrete loading square leading to the company offices. The tinning shop is above the offices and carpenter's shop. The two sheds to the right of the entrance and bridge are the fitting shop, blacksmith's shop and works' general stores. The canal basin for the works is the longish, horizontal line between the bridge and the block of offices. The canal arrives from bottom left, with the wider feeder to its right, turns right near the works' entrance and under the bridge where it vanishes from sight behind the sheds, to reappear at the Melingriffith Lock, No.44. It bends left and passes under Ty-Mawr Bridge then heads south to go under the railway bridge. The bridge at the bottom that crosses the canal and feeder is the Sunnybank Bridge. (Welsh Assembly Government Photograph Archive)

This is the canal under the works' bridge at Melingriffith in the 1930s. It was tricky for horses and boats around here, with many sharp turns and narrow towpaths leading to the loss of many horses. (S.C. Fox Photograph, Cardiff Central Library)

Melingriffith Locks in the late 1880s. This is Oak Cottage; the lock-keeper's cottage was out of shot over to the left. The famous old Melingriffith water pump was at the left in the foliage. (Author's Collection)

This old pump lifted water from the Mill Race that had done its work driving turbines and mills and was now to be lifted into the canal slightly above and near the cameraman taking this photograph. It is currently out of its foundations and is being totally overhauled by a contractor, due to be reinstalled in August 2010. (I. Jones)

After leaving the works, we go south from Whitchurch to Llandaff North. This RAF oblique of 1943 traces Ty-Mawr Bridge (6) and the railway bridge (5), then, passing the remains of the Solomon Brickworks (4), to pass under Gelli Bridge at (3). Entering Hailey Park, the canal passes what seems to be a military camp, before reaching lock 45 (1) at the Cow and Snuffers Inn, which closed down in June 2010. (2) is an attempt to show the stretches of canal where the ironworks and wharves for coal sale were. (Welsh Assembly Government Photograph Archive)

Ty-Mawr Bridge looking toward Llandaff North. (Whitchurch Library)

Primrose Hill, Llandaff North. At the middle left is Ty-Mawr House. The canal has passed under the old GWR line from Queen Street to Radyr, then Merthyr. The road to Melingriffith ran alongside the canal and you can see the portal serving that road at the right-hand side of the bridge. This boat is on maintenance duty and is probably the last boat working on the canal, in around 1944. (S.C. Fox Photograph, Whitchurch Library)

Gelli Bridge just south of the brickworks and its stack. As can be seen, there was a wharf at the left and, as the brickworks didn't have a dedicated wharf, it is likely that Gelli Wharf, which was already in existence, probably did the job for Solomon Andrews' company. The bridge was built by the GCC because the canal crossed the line of a road from Whitchurch to the Taff at Radyr, where a ford was then in regular use by people on foot and horse and cart. The church and village of Radyr were much used by the area, especially the church. The ford was at the end of Radyr Court Road. (Whitchurch Library)

The stack of the newly demolished brickworks gives the reader perspective as the canal proceeds south past Hailey Park and the rear of the houses of Hazelhurst Road, then to the rear of houses, public houses, ironworks and cottages along Station Road. (Whitchurch Library)

Llandaff Lock, No.45. The GCC's warehouse (with the white pine end) is at the middle. The famous Cow and Snuffers public house is at the right. The GCC built a large, free wharf between the warehouse and the pub. (Cardiff Central Library)

RAF overhead 1.PRU HLA/134, 1 March 1941. This shows Llandaff North and its neighbourhood. (1) is lock 46 and the bridge on College Road (at upper right); (2) is the lock house; (3) is William Beynor's house; (4) is the Three Cups public house; (5) is the site of College Ironworks; (6) is the Drill Hall; (7) is the company's warehouse; (8) is Llandaff Lock, No.45; (9) is the site of Eagle Foundry; (10) is the loading wharf for Trallwn and Merthyr; and (11) is the Cow and Snuffers. (Welsh Assembly Government Photograph Archive)

The bridge carrying the road from Llandaff to Merthyr was to become so busy that it was turnpiked in 1826. The recently closed Cow and Snuffers is the larger building further from the camera. (Whitchurch Library Collection)

College Road Canal Bridge in about 1948. The square markings in the road are the lids of holes that were to receive steel girders placed by the army in the event of invasion in 1940. The lock gates are at the left. The house is not the lock keeper's; it is William Beynon's house, built in 1813. The lock house is on the opposite bank of the canal lock 46. (Cardiff Central Library)

Bert and Bill Bladen, both of Llandaff North, are father and son with a boat bound for Pontypridd, probably flour for Hopkin Morgan's bakery. They are believed to be on the Gabalfa stretch, nearing Llandaff Yard. (Whitchurch Library)

This is James' boatyard (on the right) with a boat going upstream, empty, toward College Lock. There are two boats awaiting repair at the yard. There are three docks here for canal boats but only one of these was covered. John James had set up a boatyard at Dyffryn, on the Doctor's Canal, then he moved to Gabalfa to take over William Beynon's old boatyard. In this photograph one can see the covered dock at the right. (Whitchurch Library)

The canal at the GCC's boat maintenance yard at Gabalfa in a 1942 RAF overhead shot. To the top left one can see the Gabalfa Locks bypass water and to its right is the company engineer's house and office. In the canal, to the right of the house, are the ice-breaker and two sunken canal boats. Moving progressively to the right are: the ruin of what was the Cambrian Fuel Block Works; another of the Fuel Company's factory buildings, now Lord Bute's engineering shop; and finally the covered dock of the same company. The canal is closed for business and, as it happens, the heavy anti-aircraft battery to the south has also been through its busiest period. Western Avenue is at the extreme right and the bridge of that road is at the top left where the canal reaches the edge of the photograph. (Welsh Assembly Government Photograph Archive)

This is the northern portal of the canal bridge at Western Avenue. The canal is under housing up to this point. Any cases of subsidence along the line of the canal throughout the estate have been rectified by emergency foundation insertion. The other side of the bridge is a business park and stores. (I. Jones)

Mynachdy Lock, No.48. A boat is in the lock and is being lowered to the pound that passes the Excelsior Wire Works. The crane on the wireworks wharf can be seen in the distance. This is the north end of the lock. (Cardiff Central Library)

Mynachdy Lock is in the far distance and the Excelsior crane and wharf are at the right. The boats look abandoned and the wharf is now not used very much. Cargoes of raw material were brought up from their stores on the sea lock pound but most of the finished product went by rail from their sidings. (S.C. Fox Photograph, Welsh Industrial and Maritime Museum)

RAF overhead, 1 March 1948, 1 PRU – HLA/134. The mission took in Western Avenue at the left. The River Taff shows at the right bottom and the Excelsior Wire Works of British Ropes. The canal and the lock at Mynachdy lay between the works and Western Avenue. The Great Western Railway Cardiff to Merthyr line and the sidings for the Excelsior Works are evident. (Welsh Assembly Government Photograph Archive)

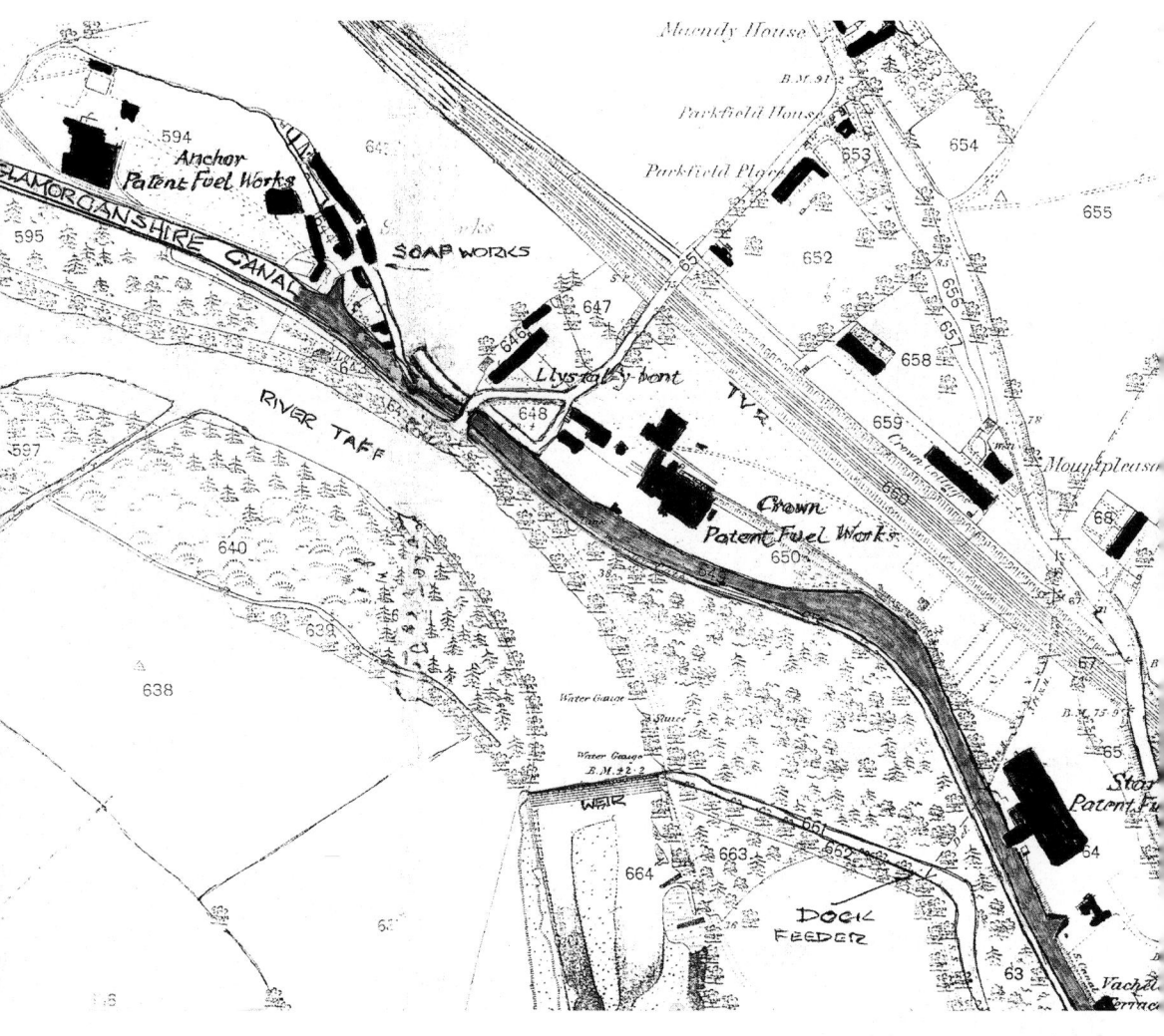

This map extract enables us to locate three fuel works and the soap works in their exact positions before the Second World War. The canal comes in at the top left from the direction of the Excelsior Works and passes the Anchor Patent Fuel Works which had its own basin off the canal. The Soap Works also used the canal for its raw materials. Then we go under the bridge at Llys-tal-y-bont, which carried the road over the railway's many lines, and reach North Road at Parkfield Place. The next place of interest as we get nearer to Cardiff is the Crown Patent Fuel Works and, a few hundred yards further, we have the Star Patent Fuel Works. All of these works were in the district of Maindy and Blackweir (named after the weir on the Taff close by that served the dock feeder). The early North Road runs vertically on the right. (Cardiff Central Library)

The Anchor Works showing the conveyors and slides used for loading the canal boats for their trip down to the West Dock at Cardiff. The photograph is taken from downstream of the works. (South Wales Coal Annual, 1913)

This boat, loaded with flour bags, probably bound for Pontypridd, has just passed under the Llys-tal-y-bont Bridge and is nearing the Anchor Works. At the bottom right is the overgrown entrance to Parry and Williams' boat dock and the Soap Works. (Cardiff Central Library)

The Star Patent Fuel Works in 1927. The canal runs along the factory's top edge with many boats tied up. North Road runs across diagonally. (*Western Mail & Echo*)

This is Blackweir Bridge as it was. It is now under the North Road car park. (Cardiff Central Library)

Showing the car park, this is the route of the canal today. At the foreground, at the cameraman's feet, was the Blackweir Bridge. (I. Jones)

Kingsway and North Road Lock, No.49, and bridge (c.1910). A full load is emerging from the lock and an empty boat is waiting to enter, going north. The GCC's boat weighbridge is in the distance on the west bank of the canal and the lock-keeper's cottage is at the east bank. The horse pulling the boat is forced to pull his head right up to the wall here to keep his barge away from the traffic on the opposite side. A few yards behind the camera the canal turns left and passes under Kingsway Bridge. (Cardiff Central Library)

Kingsway Bridge passing under North Road in 2009. The original bridge is seen under the overhanging new bridge. Go through this tunnel and you are walking along the old canal bed. (I. Jones)

The thatched cottage alongside the North Road Lock is not the lock-keeper's house but the Castle Lodge. As far as I know there was no dwelling for the lock keeper here, only an office and shelter on the roadside of the canal. In this photograph, the door facing the camera was barring the way for trespassers to the weighing machine along the bank. (*Cardiff Yesterday*)

8

College Ironworks

Situated on the north side of the canal and a few yards downstream of the lock at College Road, the works were founded by William Price in 1848 but sold on to David Davies of Merthyr and Thomas Williams of Aberdare, who turned it into a going concern in the mid-1850s. In January 1857 its two newly erected chimney stacks were blown down by a hurricane. From 1861, it operated as Lumley and Williams, the Revd Richard Lumley having taken the place of David Davies, who ceased to have an active part, having become manager of the ironworks. In 1863 Davies and Williams, together again, bought the disused Pendarren Ironworks in a vain attempt to get it working again. In the same year Thomas Williams took over the Bryncoch Brickworks and Colliery to guarantee his coal supply to College Ironworks.

College Ironworks were on stand in the late 1870s, but in 1880 the reopened works had twelve puddling furnaces and three rolling mills. Yet by 1901, when the leases expired, the works were in dereliction. Thomas Williams took over Bryncoch Brickworks and Colliery in 1863.

When the works were active, their products consisted of rolled bars of iron and wire made for use in nail-making. In its last years it was supplying plate to Llantrisant Tinplate Works. In Mineral Statistics of 1881, College Ironworks had three mills and twelve puddling furnaces.

In September 1858 the Cardiff Local Board of Health were inviting tenders for removal of the town's refuse by canal boats to College Green, probably for spreading as fertiliser, and by 1864, the area was known as the 'Llandaff nuisance' with twelve houses that doctors did not like to go to.

THE CANAL AT GABALFA LOCK NO.47

William Beynon began a boat-repairing business which developed into a boatbuilding yard just a few yards upstream of the lock. John James of Cardigan (a famous boatbuilding port) had moved to Dowlais initially but then moved to the Doctor's Canal at Dyffryn and set up a boatyard there by the bridge. He made his next move to Cardiff and Beynon's old boatyard near Gabalfa lock No.47, with his son John Rees James. There was a large house with property and a lock house for the current lock keeper on the opposite bank of the canal a hundred yards downstream. James Rees also ran a building business with his son, building several houses at Llandaff Yard.

THE EXCELSIOR WIRE ROPE WORKS

This large establishment downstream of Mynachdy Lock No.48 opened in 1929 and closed in 1966. To set up these works they brought skilled men down from the north of England to work the wire-drawing and rope-making machines. It was on the eastern side of the canal with a wharf fronting it. They owned their own boat, which brought raw materials from the sea lock pound and that part of their finished products that was to be shipped from Cardiff also went to the sea lock. But the biggest carrier of cable and wire rope went by rail on the GWR for distribution. The works became part of the British Ropes Organisation in the 1940s.

PATENT FUEL WORKS ON THE CANAL

Moving nearer to Cardiff Docks, having left Mynachdy Lock and the Excelsior Works behind, we arrive at a group of customers of the canal that stayed for a long time and were the last customers until the wharves of the sea lock.

Using small coal and pitch under pressure and heat, the combination was pressed in moulds to make blocks, each works having their own logo impressed into the finished block. Some works made small briquettes of an ovoid shape made mostly for steamship bunkering which sold very well.

The trade was first developed in France in 1842 and the first British plant was set up in Newcastle. Wales, with its many collieries, soon followed and works were opened in Swansea in 1847. Cardiff followed in 1857, on the bank of the Glamorganshire Canal. The small coal was readily available from every colliery screen or at the dockside, where the coal tips and hoists were crashing coal into the chutes. Steam coal was best for the patent fuel and there was more of that at Cardiff than anywhere else. The pitch too was available as it was a by-product of gas works and coke works. By the late 1890s over 90 per cent of South Wales' patent fuel was exported for use by the world's navies and railway companies. As far as the Glamorganshire Canal was concerned, four factories were established by rival companies on its banks in north Cardiff, another at Taffs Well and two others on the Aberdare Canal. The first of these was the Cardiff Preserved Coal and Coke Co. Ltd in June 1857, just below Lystalybont Bridge. It was founded by Henry Walker Wood of Briton Ferry, who had been operating a fuel works at Port Talbot since 1851. Within three years the company needed financial restructuring and was sold to new shareholders operating as the Crown Preserved Coal and Coke Co. Ltd. Wood retained a minority holding of shares and his son T.H. Wood remained as manager.

The Anchor Patent Fuel Works opened in 1866 nearby. It was a French company owned by Tinel & Co., who had been purchasing most of their small coal from Cardiff for their works at Le Havre. The Anchor Works were managed by Louis Jean Baptiste Gueret from 1868 to 1879, when the firm relinquished the Cardiff patent fuel business solely to Louis and his brother Henri. When Henri died in 1895 the Anchor was formed into the limited company of L. Gueret Ltd, and in 1909 he was operating eight other patent fuel works in France. This company purchased a considerable share of the Albion Colliery at Cilfynydd.

The engineer of the Crown Patent Fuel Works, T.P. Balls, left Crown to install his own patents for fuel manufacture at the Cwmbach Works on the Aberdare Canal in 1872 and his former boss, Mr Heath, left the Crown Works to start his own works to use his own ideas of production – the Star Patent Fuel Works with his partner Tom Evans, who was still engineer at Werfa Colliery in Aberdare. The Star Works were erected on the site of the old Maindy Forge and Foundry just down the canal from the Crown Works.

In 1875 the Cambrian Patent Fuel Works too were established just north of Mynachdy at Gabalfa on a three-acre site with a 1,165ft frontage to the canal, so by the mid-1870s there were three patent fuel works operating within a mile of each other in the area previously known as Maindy, Lystalybont and Blackweir, and another less than a mile from them. They all relied on the canal for their raw materials and for sending their finished product to Cardiff. The products, by now, had to be shipped from the Bute Docks, not the sea lock pound. This was due to a clause in the leases at Bute lands to build their works. All traffic was to go down to the canal then pass through the Junction Canal into the West Dock.

To meet the demand for collection of small coal and pitch, plus the delivery of fuel blocks to the docks, a new fleet of canal boats was built. Such was the demand that at least one boat builder was able to concentrate solely on the supply and maintenance of fuel boats for the canal.

Crown went into liquidation for the second time in 1877 and the works were sold to William Butler of Bristol, who placed the management of it in the hands of his son Samuel. They remodelled the works and adopted their own patents, substituting steam pressure with hydraulic. A railway siding was put into the Taff Vale Railway so that coal could be brought in by rail, whilst retaining the canal route to the docks. Butler also ran a fleet of barges and tugs collecting tar from various places on the River Severn, the Avon at Bristol and the Kennet and Avon Canal. This was brought back to their works at the Parting in Gloucester and Crews Hole at Bristol for refining into pitch. The main destination for the pitch was the Crown Patent Fuel Works.

The Anchor Works had been the first to connect to the railway a few years before. By the turn of the century the Star too had rail access. Only the Cambrian relied totally on the canal for transport. At the end of 1882 the Cambrian Works were recapitalised as the Cambrian Patent Fuel Co. Ltd and put in their best year by shipping 53,634 tons through the Junction Lock; for the first time, this was more than the Star at 53,305 tons and the Anchor at 42,616 tons. However, by 1893 the company had ceased production. Being totally reliant on the canal for transport, it may never have recovered from the eight-month sea lock pound closure of 1886 when lock 51 was being put in. In the same year, the Crown Fuel Works took advantage of the building of the Roath Dock and moved in near the Roath basin with a brand new factory at the dockside with a capacity to produce 3,000 tons a week.

Crown's old plant was also still in production on the canal side as there was so much demand. Indeed, as the new plant on the docks completed its first year, the production figure was 7,500 tons per week. It was an economic success as the required materials, coal and pitch, were at hand and the finished blocks were loaded directly to ship. Crown built a copy of this dockside factory at the new dock at Port Talbot and, once these new works were up and running, the old Crown factory at Maindy was closed. This Maindy site became the Hall, Lewis & Co. Wagon Works, which, by 1935, were part of the Powell Duffryn Associated Collieries – Cambrian Wagon Works. This company was building, repairing and leasing railway wagons. The Crown Company also had a depot, canal-side, at the northern end of Harrowby Street with a branch railway line from the Great Western Railway Clarence Road, Riverside line.

Just as for steam coal proper, the peak period for the fuel works was the decade before the outbreak of the Great War. The Star Works at that time was producing about 200,000 tons of fuel annually and the total output for the Cardiff trio – Star, Anchor and Crown – peaked in 1913 at 716,899 tons. After the war, and the depression that followed in shipping because of reduced sales, plus oil gradually replacing coal as a fuel for ships, the Star Company closed in December 1927 and their assets were taken over by British Briquettes.

In 1919 the assets of the Anchor Works were transferred to a new L. Gueret & Co. Ltd, but this was not a fuel block trade. This company became Gueret, Llewellyn and Merrett Ltd, before it was absorbed into British Briquettes in 1929 and closed down. British Briquettes was formed in 1929 to effect an amalgamation of the fuel block industry in South Wales, taking over Crown Preserved Coal Co. Ltd, Rose Patent Fuel Co. Ltd, Graigola Merthyr Co. Ltd, Pacific Fuel Co. Ltd, Star Patent Fuel Co. Ltd, Gueret, Llewellyn and Merret Ltd, Arrow Fuel Co. Ltd and all the issued shares of Abertillery Pitch and Benzole Co. Ltd. This new company had a wide range of activities other than patent fuel and in 1936 was absorbed into Powell Duffryn Associated Collieries Ltd, but, by that time, it was six years since patent fuel had been carried on the Glamorganshire and Aberdare canals.

This fine photograph shows the squalid parts of the boatman's day, at Crockherbtown Lock, No.50. The wall on the left has the friary on the other side. Queen Street is ahead on the other side of the white display window box and the Carlton Restaurant is to the right. The canal descends steeply under the arch then the boat is pulled by chains through a long tunnel that passes well under the width of Queen Street. The horse is led up to street level. (Ian L. Wright)

On passing through the tunnel, we reach the south portal. Above is Tunnel Court which was demolished in 1959. All this is now under the St David's Centre, but I am told that the tunnel is extant and access is possible with the right key! Note the Corporation official on the wharf. (Cardiff Central Library Local Studies)

It is June 1957 and they are filling in the canal where it passed behind Working Street. Hill's Terrace is on the right and the tunnel under Queen's Street is ahead in the shadows. All the warehouses in this area were built in the 1920s and were demolished in 1978. The canal now curves to the west and we pass under Hayes Bridge. (Cardiff Central Library Local Studies)

After clearing Hayes Bridge, we head more to the west and follow Mill Lane down as far as the Monument. This expression of 'the Monument' was common from before Second World War because it was the key to navigation of Cardiff streets. It was a terminus for trams and buses; it was near the station; it was a crossroads that led to Penarth: and, in the middle of the square, was the Monument. Facing St Mary Street, it was a high square plinth with a high statue of the 2nd Marquess of Bute. It was irrelevant to most people who the statue represented but if you needed to meet anyone, this was the place to do so. So, as we approach the monument along Mill Lane we pass under the Custom House Street Bridge and turn sharp left into the final stretch of the Glamorganshire Canal as it then passed beneath the railway bridge carrying the main line to London. (Cardiff Central Library Local Studies)

The canal boat has reached the last stretch and we are facing south toward the sea lock pound away in the distance. The wharves and warehouses now appear. The GCC's wharf is on the right and the Custom House is at the left. Note the many boats moored, it must be Sunday! (*Western Mail & Echo*)

Looking north, the Monument is at the left of the photograph. St Mary Street facing us at the top left is decorated for reasons unknown. The GCC's wharf is at the left bottom and the Mill Lane exit is at the right of the bridge. (*Western Mail & Echo*)

This is the cut that joined the canal with Bute West Dock, called the Junction Canal; it had a similar connection to the East Dock when that was built. Cargoes from the fuel works no longer loaded in the canal. It was by this time mostly large ocean-going ships that took these cargoes by conveyors at West Dock. The gap left between the two large abutments once carried railway lines. (Glamorgan Record Office)

Fuel blocks being loaded intensively at West Dock, after the boats bringing these down from Maindy have left the canal above lock 51 to discharge here. (*Cardiff Times*)

The last stop on the Glamorganshire Canal. This photograph was taken when new huge balance beams were fitted to the lock gates at Sea Lock, in 1891. The ships in the sea lock pound looked increasingly tiny as the big stuff came into the East and West Docks. (Cardiff Library)

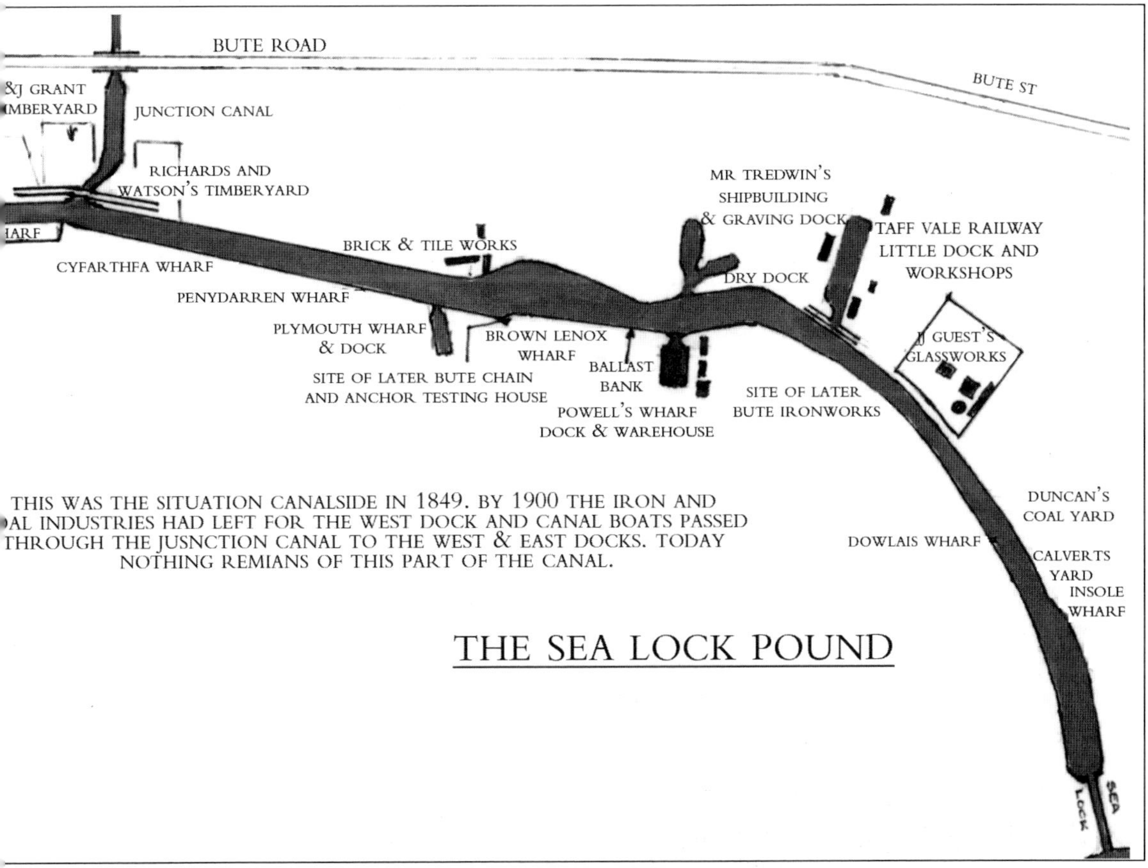

BUTE ROAD

BUTE ST

&J GRANT
IMBERYARD

JUNCTION CANAL

RICHARDS AND
WATSON'S TIMBERYARD

MR TREDWIN'S
SHIPBUILDING
& GRAVING DOCK

TAFF VALE RAILWAY
LITTLE DOCK AND
WORKSHOPS

IARF

BRICK & TILE WORKS

CYFARTHFA WHARF

PENYDARREN WHARF

DRY DOCK

PLYMOUTH WHARF
& DOCK

BROWN LENOX
WHARF

BALLAST
BANK

SITE OF LATER BUTE CHAIN
AND ANCHOR TESTING HOUSE

POWELL'S WHARF
DOCK & WAREHOUSE

SITE OF LATER
BUTE IRONWORKS

GUEST'S
GLASSWORKS

THIS WAS THE SITUATION CANALSIDE IN 1849. BY 1900 THE IRON AND
AL INDUSTRIES HAD LEFT FOR THE WEST DOCK AND CANAL BOATS PASSED
THROUGH THE JUSNCTION CANAL TO THE WEST & EAST DOCKS. TODAY
NOTHING REMIANS OF THIS PART OF THE CANAL.

DUNCAN'S
COAL YARD

DOWLAIS WHARF

CALVERTS
YARD
INSOLE
WHARF

THE SEA LOCK POUND

SEA LOCK

This map describes the final length of the canal. The boat owners and coal owners mixed with ironmasters to get their produce out of this tiny dock to make money all over the map. What a busy place it must have been! (I. Jones)

The sea lock pound in April 1943. The vessel is the *Catherine Ethel* of the Sandridge Company. The new Sea Lock Hotel is alongside the ship, which was, unfortunately, the culprit on 5 December 1951 when she was accidentally driven into the sea lock gate and emptied the canal of all its water, never to be filled again! The Cardiff Corporation had long desired to close this old-fashioned waterway down and had no intention of rectifying this calamity. (Cardiff Central Library)

This RAF oblique dates from 8 April 1942, when the canal was closed down, except that part between James Street Bridge and the sea lock, which one can see at the bottom left. James Street Bridge is a short distance back up the canal. The white dot at the north end near the railway bridge is marking the entrance to the Junction Canal at West Dock. Note the barrage balloon over the important commercial area of the docks. At the top left of the canal, one can see the timber float. In the middle of this photograph, between the River Taff and the canal, can be seen the long sheds, running west to east, of the massive Curran's Works, forgotten now but a thriving place during the war. (Welsh Assembly Government Photographic Archive)

9

The Aberdare Canal

The Glamorganshire Canal had proved its worth in bringing iron down to Cardiff and there were good reasons for building a similar canal down the Cynon Valley to meet the Glamorganshire at Navigation (Abercynon), but there were not the number of customer iron companies to fund it.

Hirwaun was the first in the area in 1757, when John Maybery leased land there to build a furnace. A new lease was given to John Wilkins, John Maybery and Mary Maybery in 1760. These partners disposed of the Hirwaun Ironworks to John Wasse of Stratford and William King, a Bristol glassmaker. King died insolvent and the lease, that was due to run for fourteen years, failed. After legal problems the lease was terminated in 1777.

In 1780, the concern was leased to Anthony Bacon at a cost of £133 6s 8d annually. Bacon put the works into repair and continued the production of iron on the site until his death in 1786. Hirwaun was then left to his sons Anthony and Thomas Bacon and, as they were minors, the property was placed under the custody of the Court of Chancery. The court granted the lease of the works to Samuel Glover, who was allowed to carry on iron making on the site during the minority of the sons. When they came of age, they had little interest in the business and, in 1803, they demised the property to Francis William Bowzer of Hendon, Simon Oliver of Bristol, Lionel Oliver of Bristol and Jeremiah Homfray of Llandaff. Within a few years, Simon Oliver retired from the partnership and his place was taken by George Overton. In 1805 there was still only one furnace and it produced 450 tons of iron. After this date money was found to develop the site because, in 1813, when the Hirwaun Ironworks were up for sale, it consisted of two furnaces, two cast houses, an air furnace, two fineries, a blast engine, a forge, ten puddling furnaces, five balling furnaces, carpenter's and smith's shops, a lathe, three limekilns, four collieries, iron ore levels, cottages and tenements.

The iron trade was in deep depression at this time and a sale was not concluded, with the result that, in the following year, the partnership went bankrupt in 1814. The ironworks remained unoccupied until William Crawshay of Cyfarthfa took over the lease in 1819. In 1791 the Neath Canal was authorised from the top of which, at Glynneath, it was practicable to build a tramroad to Hirwaun and the pressure was put on to Hirwaun's iron to the Neath Canal. Meanwhile, in 1792, there had been a subscription among a group of those whose interests lay in the other direction, to the east and the Glamorganshire Canal. The subscription was to pay for a survey for a canal from the Glamorganshire Canal to the Neath Canal. This was altered and in 1793 the Aberdare Canal was authorised from Abercynon to Tydraw, which went across the Cynon Valley from the village of Aberdare. Another clause in the authorisation states:

> And for making a railway, or stone road, from thence to Abernant, as well as other tramroads within 8 miles of its line.

The Aberdare turnpike road from Abercynon, through Aberdare, to Abernant (Glynneath) was authorised on the same day.

The Company's capital was £22,300 with power to raise £11,000 more if necessary. A total of £22,100 was subscribed, some by Samuel and Jeremiah Homfray of Penydarren, Richard Hill of Plymouth Works and John Partridge – soon to be concerned with Melingriffith – who together put up £3,500. Much came also from Sam Glover and other men from Birmingham and Tamworth, who put up £3,200. Two local men, Hugh Lord of Aberaman and John Knight of Duffryn, put up £1,000 each. Brecon interests, notably the Wilkins of the Old Bank who were property owners around Hirwaun, the Powell's £7,600, and the four Dadfords, all concerned with Welsh canals in one way or another, reached a total of £2,000.

The list of trustees of the new road included Glover, the Wilkins, the Homfrays, Hill, Partridge, Richard Crawshay, Thomas Guest and Williams Taitt.

At this time there was just Hirwaun Ironworks and the canal committee decided that it would be best if cutting the canal was postponed until iron making had spread and more goods would be available to support the intended canal, but tramroad building could go ahead. The valley, at this period in time, had no roads, just a track. It was open country with isolated farms and even if a canal was cut, it would have been difficult to bring provisions, stone and timber up. It was decided to build a tramroad from the proposed canal basin at Aberdare to join Glover's railroad at Hirwaun Common and from thence to the lime rock at Penderyn. This, and several other tramroads, was built by the Aberdare Canal Company (ACC). The turnpike road was begun in May 1803 and finished in 1810.

The Neath Canal Company was now making moves to build a tramroad through Hirwaun to Aberdare to get the traffic from that area.

In 1800 the formation of a partnership to lease land and build the Llwydcoed Ironworks (also known as the Aberdare Ironworks) was formed. In that year Samuel Glover leased land at Aberdare to John Thompson, John Hodgett, and George and John Scale. The lease was to run for seventy years with a payment of £1,000 per annum. Two furnaces were built and in 1805 they produced 3,586 tons of iron. In 1811 the partnership was William Thompson, George Scale, John Scale, Samuel Homfray and William Forman. The company expanded considerably when, in 1819, they purchased the neighbouring Abernant Ironworks. The Abernant property had been leased in 1801 for ninety-nine years to Jeremiah Homfray and James Birch. Three furnaces were built by 1807. They enticed fresh money into the concern, for in

1802, James Tappenden of Stourmouth, and Kent and Francis Tappenden of London joined the partnership.

Homfray and Birch left the company in 1807 after a disagreement with the Tappendens. The Tappendens were involved in a costly court action against the Neath Canal Company. On losing this action in 1814, they were forced into bankruptcy. The Abernant Works were later sold to the Aberdare Iron Company in 1819.

In 1823 the company was operating three furnaces at each site (Abernant and Aberdare) and 5,676 tons of iron was being produced, rising to 11,440 tons in 1826 and 12,571 tons in 1830. At the end of the 1830s the Aberdare Iron Company was producing 350–400 tons of iron per week.

After a dispute between Henry and Mary Scale on one side and Rowland Fothergill on the other in 1846 at the Court of Chancery, the company was sold and purchased by a new company headed by the Fothergills. The Aberdare Iron Company was managed by Rowland Fothergill, and later by his nephew Richard Fothergill. Puddling furnaces were added and, under Richard Fothergill, both works were modernised.

In August 1809 it was agreed to build the canal, and a tramroad, from it to the Aberdare Ironworks at Llwydcoed. Work began in 1810 with Thomas Sheasby as resident engineer. Control of the canal had changed after this long delay since 1793 when the canal committee had formed. The Homfrays had long ago sold their shares and Richard Hill had left the committee in 1809 to join the Wilkins interest. There now joined the group: John and George Scale of Aberdare Works, Francis Tappenden, F.W. Bowzer, Overton and Oliver of Hirwaun and Joseph Bailey, with colliery interests. They ran the canal for ten years. In August 1811 Sheasby resigned and went as clerk and engineer to the Severn and Wye Company, George Overton taking over from him.

The canal was opened in mid-August 1812. It was 6¾ miles long with two locks, at Cwmbach and Dyffryn, and a stop lock at the junction. The boats were of the same size as the Glamorganshire Canal boats. A tramroad connection from the canal head had been built about 1811 to join the Glynneath–Hirwaun–Abernant tramroad at Llwydcoed. A private branch from the Gladlys Ironworks later joined this at Trecynon, whilst another from the Abernant Ironworks to the canal head was probably built after 1819, when they were taken over by the owners of Aberdare Ironworks. The cost of the canal at opening was approximately £26,220 which was raised by making a share worth £120. Then just as the canal was opened, the trade in iron took a dive that took three years to reverse.

William Crawshay and his son William Crawshay II had brought Hirwaun Works back to production and these two, along with Crawshay's henchman Thomas Charles and the coal owner Dr Richard Griffiths, became shareholders of the ACC. They also joined the ACC committee along with another group, including George Scale, J.B. Bruce and Richard Fothergill. In 1820 the ACC agreed to proposals made by these two groups, which reduced toll charges on the ACC by half for three years if the companies used the canal and tramroads exclusively for their products, which meant sending their goods to the GCC and thence to Cardiff, instead of via the Neath Canal to Neath or Swansea. This seems to have solved the Neath challenge as the canal company began its busiest years.

In 1882 William Crawshay began to emulate his grandfather and attempted to be the major shareholder of the ACC. He had already increased his share from twenty-five to fifty-five of the 225 total, and seems to have been behind most of the stimulating proposals made. In 1825 it was resolved:

> The canal being now in an efficient state for carrying down the increased trade that the tramroads must now be put in a state of thorough repair.

This was because the company had recently spent most of its resources on raising the banks of the canal to provide greater depth, so that the boats could carry their full 25 tons.

The traffic on the canal in 1828 was still only 59,525 tons, but the Crawshays had started further ironworks at Penderyn and iron and tinplate works at Treforest, which used iron carried from Hirwaun. The Gladlys Ironworks were built in Aberdare in 1827 with a single furnace. The partners in this concern were George Rowland Morgan, Edward Morgan Williams and Matthew Wayne.

These works were advertised for sale in 1835 and consisted of 350 acres of mineral property, with the works employing 150 people. The single furnace was capable of producing 1,700 to 2,000 tons of iron annually, blown by an engine which also supplied the blast to a refinery. Other property on the site included a cast house, stove room, turning room, cupola, smith's shop, weighing machine, bridge house, carpenter's shop, office, punching machine and turning lathe. There were also three calcining kilns for burning off impurities in the iron ore. The iron produced at Gladlys was praised for its strength in castings and for engine parts. The sale of the works was not made and remained in the hands of a company headed by Matthew Wayne and, after his death, was run by his son Thomas Wayne.

A connection to the ACC for its iron was made by a tramroad that ran to Trecynon where it joined the Glynneath–Hirwaun–Abernant tramroad. Another later development in the iron industry of this valley was the Aberaman mineral estate purchased by Crawshay Bailey in 1837 but he did not exploit this investment until 1845, when he constructed three furnaces on the property and iron was made. The furnaces were out of blast in 1854, they were continually at work from 1855 to 1866.

There was an attempted sale of the works in 1862 for £250,000, followed by another reported sale in 1864 when a £100,000 deposit was paid and the Aberaman Iron Company was floated. But there was something wrong. The affairs of the Aberaman Iron Company were wound up by the Court of Chancery in 1867 and it is that year that the ironworks were taken over by the Powell Dyffryn Steam Coal Company. It would seem that the Aberaman Works were acquired for their valuable mineral ground, which continues to be exploited while the works remained unoccupied.

In 1841, after two years of negotiation between Crawshay and Fothergill (as the principal freighters on the ACC), the company bought from Benjamin Hall (who was the assignee of the Abernant Works Estate when the Tappendens went bankrupt) the Rhigos–Abernant portion of the old Glynneath–Abernant tramroad, part of which formed the connection to the Hirwaun–Penderyn section of the company's tramroad and the Llwydcoed section. The money was raised by loan and the Hirwaun–Rhigos section was allowed to decay since it was agreed that traffic should not pass that way.

There was a changing situation on the ACC as coal was supplementing the iron as cargoes. The steam coal demand was growing and, in 1842, Thomas Powell was asking permission to carry coal from his famous colliery in containers in the boats. Sixteen steam coal pits were sunk in the valley between 1840 and 1853. Dividends rose and in 1843 the swelling trade was helped by a 25 per cent reduction in the tolls on coal, iron, iron ore, ironstone, limestone, pit wood and quarry stone. Indeed, water to carry the trade was now getting short and a pumping engine was put to work in 1846, bringing water from the Cynon. The cost was £4,143, two-thirds of which was paid by the GCC as the water would arrive on their canal eventually.

By 1848 the tonnage had trebled in ten years to 159,633 and the dividends had equally improved. Before then, competition had appeared. In July 1845 the Aberdare Railway had been incorporated. it was to run from the Taff Vale Railway, to a certain tramroad leading from the Hirwaun Ironworks to the Aberdare Canal. With Sir Josiah J. Guest and Crawshay Bailey among its directors, it was opened in May 1846 and leased, in perpetuity, from the beginning of 1847 to the Taff Vale Railway. This competition caused an immediate drop in the average dividends on the canal as many of the coal levels and pits built sidings to the railway.

Another railway was interested in the steam coal traffic. The Vale of Neath had reached the town of Aberdare in 1851 and it had considered buying some of the old tramroads, but Crawshay did not think it was of profit to the canal company and it was dropped.

A junction between the railway at Aberdare and the canal head was the favoured option for the Vale of Neath to get in on Crawshay's downward traffic and coal shipments from Dyffryn and collieries further south for carriage to Swansea and Neath. After an abortive attempt to load coal at the canal head with little space to load and unload the boats, the Vale of Neath dropped the wharf at the head and made an extension that took them alongside the canal, south, to the middle Dyffryn Colliery, and it opened in 1856. This extension was named the Aberdare Valley Line. It was leased to the Vale of Neath Railway and it became their property in 1864.

In 1864, another competitor to the canal appeared in the form of the Great Western Railway and they built a line from Pontypool Road to Mountain Ash, which made inroads into the profits of the ACC.

Almost all collieries had made a connection with one of the railways but, because of sheer volume of coal and iron to be transported, one way or another the Aberdare Canal Company's tonnage actually improved.

The 1850s and 1860s were to be the most prosperous years for the canal. For example, we can look at David Davies's Blaengwawr Colliery for this period. In March 1855, 111 boatloads of large coal amounting to 2,314 tons 18cwt went by canal. The colliery was connected to the railway in March 1856 and this stimulated production. As the months passed by the railway was able to take up the colliery's increase in output whilst the quantity shipped by canal did fall. In a sample month – August 1859 – 7,947 tons 17cwt went by rail and only 885 tons 15cwt by canal.

The GCC tonnage book for down traffic in August 1863 shows the decline of iron and the flourishing coal being transported at the Aberdare Canal's junction with the Glamorganshire Canal at lock No.17:

Aberdare Iron Company	Iron
Lletty Shenkin Coal Company	Coal
Aberdare Coal Company	Coal
David Davies	Coal
George Insole and Son	Coal
Thomas Powell and Son	Coal
Nixon Taylor and Company	Coal
Morgan Edwards and Company	Coal
Crown Preserved Coal Company	Coal
Chivers, Todd, Chivers	Coal
Evan Williams	Coal
William Morgan	Coal
George Steel	Coal
Brown Lenox and Company	Coal
Iron Company	Iron
William John	Coal

The population of Aberdare was only 4,000 in 1831, but had expanded to 32,000 by 1861. One of the functions of both railways and canals was to bring provisions up for the shops of the new mining communities. In 1850, for instance, we find the canal carrier Evan Griffiths trading regularly between Cardiff Wharf and Aberdare and the barges of the Hirwaun Boating Company, providing 'conveyance by water' to Aberdare three times per week. But things were not to go so well in the future because the canal was in decline. By 1872, thirteen coal trains a day were being moved eastwards. By 1885 the number of coal trains was forty daily. In 1858 the ACC receipts stood at a healthy £4,352. By 1868 they had fallen to £2,837.

The strike and stoppage at the Aberdare Ironworks in 1875 then produced a desperate situation for the canal company. But worst of all, from the end of the 1860s, the canal was in trouble with mining subsidence and the company's efforts to get compensation from the larger coal owners were unsuccessful. Burnefeat & Co., who had taken over the Lletty Shenkin Colliery in 1872, paid a paltry £23 7s 5d but the main culprits were Powell Dyffryn & Co., which had been formed on the death of Thomas Powell.

In 1885 the Marquess of Bute acquired the Glamorganshire and Aberdare canals by buying the shares held by the Crawshay family. Bute himself became chairman. He wanted to have the water used by the canals for his docks at Cardiff, yet he merged the two canals and spent much money in improving those parts that were still in use. In 1887 the company began the regular carrying of merchandise between Cardiff,

Pontypridd and Merthyr on the Glamorganshire, and, on the Aberdare Canal, between Cardiff, Mountain Ash and Aberdare. Bute's efforts at Aberdare were unsuccessful as the railway competition was overwhelming.

The mining subsidence was causing sinking towpaths and subsiding bridges. That and the wretched canal traffic caused the tonnage to dwindle to only 7,885 tons in 1897 and, in that year, the annual engineering bill for stopping leaks and raising flooded towpaths rose to £46 per mile. In 1900 the canal had become unworkable and the decision to close it on the grounds of public safety was made.

It was bought by the Urban District Councils of Aberdare and Mountain Ash in 1923, which was soon followed by the conversion of the canal into a road up the valley, designated the A4059 and B4275. For this reason few signs of the canal remain, except in the conserved area between Ynyscynon Bridge and the canal head.

The old Aberdare Canal Company finally went into liquidation in 1955.

This was the Canal Head House at Cwmbach, Aberdare in 1890. (Welsh Assembly Government Photographic Archive)

After a transformation by new owners, Canal Head House in 2010. (I. Jones)

The Aberdare Canal Company's wharf and works' house at the canal head, just after closure of the canal. (Welsh Assembly Government Photographic Archive)

Scales Bridge at Cwmbach on the abandoned Aberdare Canal in 1925. (Rhondda-Cynon-Taff Libraries)

This is the bridge that was built for the Abernant Railway of the Abernant Iron Company. It is a skew bridge and it carried minerals to join the Taff Vale Railway's Aberdare Railway. Before the railway line reached here, it had already left a spur that brought a track down to the north side of the canal to a wharf for shipments on the canal to Cardiff. The bridge is made of cast and wrought iron and encased in concrete. It is only wide enough for one track. It is now suffering from 100 years of neglect, with trees and bushes growing inside. (I. Jones)

The restoration of this length of canal was achieved by local volunteers. The turning off to the left of the photograph was the place where the Blaengwawr tramroad to the dock emerged. (I. Jones)

Right: This is the Cwmbach to Llwydcoed rail bed today. (I. Jones)

Below: The Werfa Railway line, running from the tunnel down to the canal. (I. Jones)

Below: The tunnel belongs to the Werfa Railway, once a tramroad. Built by the Werfa Colliery Company, it travelled from the pit on the long straight run downhill to the canal. It was built before the Cwmbach to Llwydcoed Railway, which crosses over the Werfa line, so the bridge was of Llwydcoed making. (I. Jones)

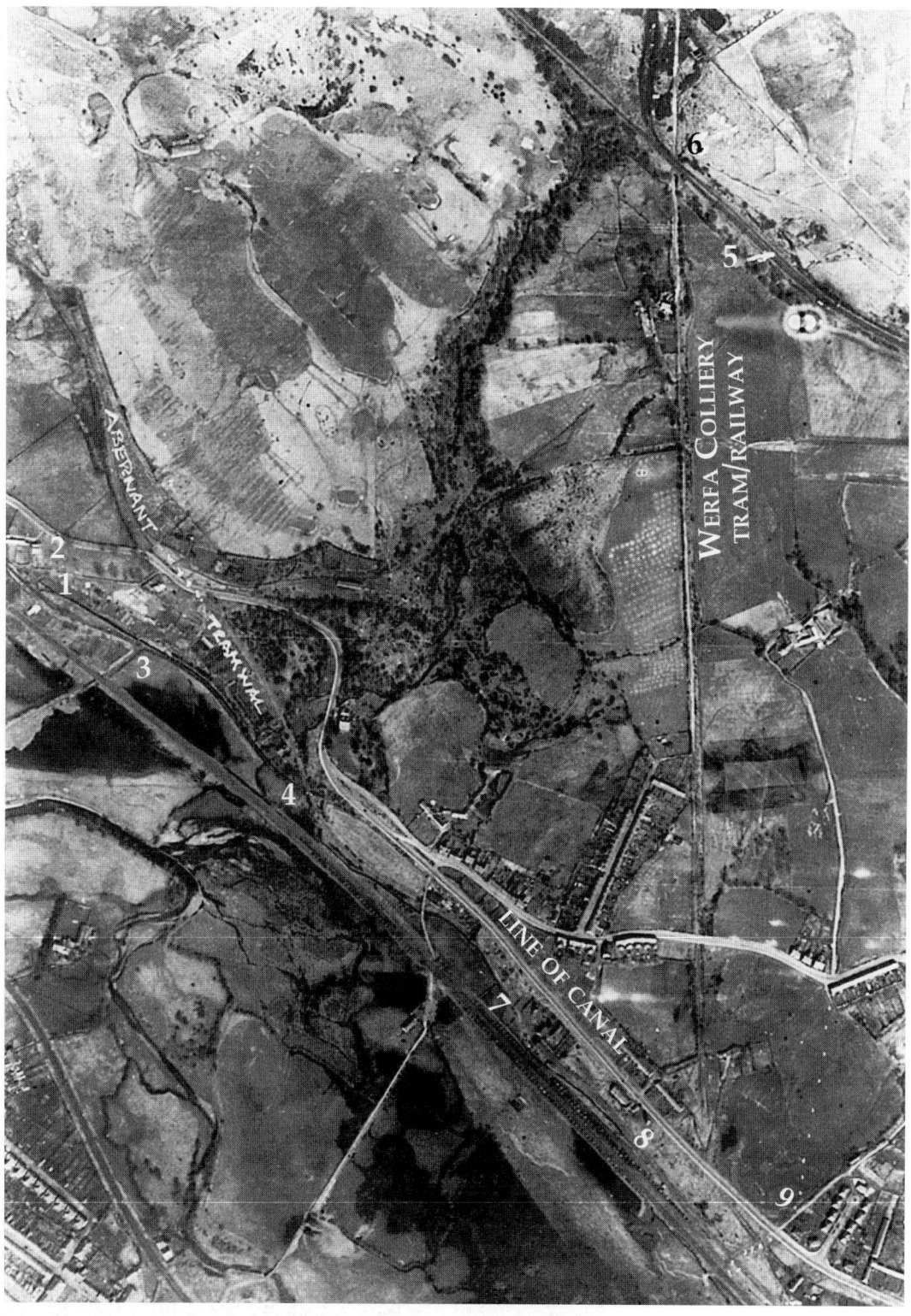

The Aberdare Canal head in Cwmbach, April 1947: (1) is the canal head; (2) is Canal Head House; (3) is Blaengwawr; (4) is Abernant Railway Bridge; (5) is the Cwmbach to Llwydcoed Railway; (6) is the bridge over the Werfa Colliery Railway; (7) is the remains of High Dyffryn Colliery; (8) is the remains of Cwmbach little pit; and (9) is the Cwmbach New pit. (Welsh Assembly Government Photographic Archive)

This old photograph, 'Frost and Fog', taken by Mr Richard Berry in 1880, shows a body of men of Aberdare in Sunday dress, defying the cold and daring the ice to break. The canal bridge is called 'Cae Draw Nant' and is near Lord Aberdare's residence – Dyffryn House. (Welsh Assembly Government Photographic Archive)

It is 1920 and the Aberdare Canal has been filled in with colliery waste. As in all the RAF photographs, we can only follow the line of the canal that is left, until roads covered them: (1) is the railway and bridge to Lower Dyffryn and George Colliery, and also marks the site of Powell basin; (2) is High Dyffryn Colliery; (3) is Dyffryn Lock; and (4) is Cae Draw Nant Bridge and Aqueduct. (Welsh Assembly Government Photographic Archive)

This 2010 photo simply portrays the way the canal is vanishing from our landscape. These are the remains of locks 16 and 17 now in a private garden, where it is looked after very well. The Aberdare Canal joined from the left (west) on this staircase of locks. (I. Jones)

The junction of the Aberdare Canal with the Glamorganshire Canal at Abercynon. At the top left of the photograph, the Glamorganshire Canal flows south to (1) Pont Haiarn and then passes under (2) the locks 14 and 15 then on to (3) locks 16 and 17. At the space between (3) and (4), Lock-y-Waun, the Aberdare Canal joins from the left (west) where there is a boat dock. The combined canals pass through Lock-y-Waun and continue the rest of this remarkable staircase of locks down the hillside to (5) Lock Odyn Galch. Then we pass on to (6) locks 22 and 23, Lock Stackhouse. Then we travel on down to (7) Lock Isaf and (8) the aqueduct, which later became a road bridge, and finally (9) the Glamorganshire Canal's basin. (Welsh Assembly Government Photographic Archive)

Right: In this RAF image from April 1947, Penrhiwceiber Town is at the upper left but the dark shadow of the mountain of Mynydd Merthyr darkens much of this photograph. The snake-like white strip is the Aberdare Canal. At the top of the photograph is Cwmcynon Colliery (left) and Mountain Ash is above the photograph. (1) is the site of Letty Turner Bridge; and (2) is the Tunnel Bridge site. (Welsh Assembly Government Photographic Archive)

Left: This oblique of Mountain Ash was taken on 7 March 1945 and shows the line of the dry bed of the Aberdare Canal, running along the north-western side of the town, from the bottom left to upper right of the photograph. At the left, it has passed Dyffryn Lock and Deep Dyffryn Colliery. Then we travel past Ffrwd Bridge and on through the town, always at the foot of the mountain. Nixon's Navigation Colliery is on the western side, opposite. The Colliery in the distance is that of Cwmcynon. Coal spoil is tipped on both sides of the canal, all the way down the valley. The line of the canal was later to become the new road, the A4059. (Welsh Assembly Government Photographic Archive)

Epilogue

So we have arrived at the end of this story, brief as it is. It is not, however, the end of interest in the canal, as evidenced by numerous letters to the local newspapers and on the internet, where people, some too young to have seen it for themselves, are complaining of its closure when it could be one of the amenities of modern Cardiff and District.

This waterway, built 220 years ago, did have an interesting touch to it and it can best be realised by the time and money that has been spent on overhauling – indeed more like rebuilding at great expense – the water pump at Melingriffith. This project has been made possible by the joint efforts of the Friends of Melingriffith Water Pump, Oxford House (Risca) Industrial Society, the Inland Waterways Society, South Glamorgan County Council, Cardiff City Council, and the Welsh Assembly Government, through CADW working with the contractor Pen-y-Bryn Engineering Ltd.

This pump was built following court action in a dispute between the Melingriffith Tin Plate Works and the Glamorganshire Canal Company. The result was that water that had flowed through the Melingriffith Mill's watermills and turbines and was now free to return to the River Taff would, in future, be pumped into the canal to bring its water level up.

The contractor removed 300 tons of rubble from the watercourse, replaced the waterwheel shaft, all the paddle wheel blades and fixings, the steel work that supported the pump cylinders, and the 10in x 10in oak A-frames, reconstructed the stone walls of the dam and sluices, and reinstated the water supply. The job began in the summer of 2010 and is due to run (in the presence of many onlookers) to 1 July 2011. It will be turned by electric motor, not water.

Left: Summer 2010: The stripping-down of the Melingriffith Water Pump and preparation to transport it to Pen-y-Bryn Engineering Ltd. *Centre:* Summer 2011 and the overhauled pump is ready for its first run on the evening of 1 July. *Right:* Showing the rocking beams and the A-frames. The two pumping cylinders are at the right.